INTRODUCTION TO BIOCHEMISTRY

SECOND EDITION

 SAUNDERS GOLDEN SUNBURST SERIES

JOSEPH I. ROUTH

Department of Biochemistry
School of Medicine, University of Iowa

SAUNDERS COLLEGE

Philadelphia

W. B. Saunders Company: West Washington Square
Philadelphia, PA 19105

1 St. Anne's Road
Eastbourne, East Sussex BN21 3UN, England

1 Goldthorne Avenue
Toronto, Ontario M8Z 5T9, Canada

Library of Congress Cataloging in Publication Data

Routh, Joseph Isaac, 1910–
  Introduction to biochemistry.

  (Saunders golden sunburst series)
  Bibliography: p.
  Includes index.
  1. Biological chemistry. I. Title.
QP514.2.R67 1977   574.1′92   76-28945
ISBN 0-7216-7759-2

Introduction to Biochemistry   ISBN   0-7216-7759-2

456   026   10 9 8 7 6 5

PREFACE

The preparation of a new edition of a limited-size text covering the fundamental concepts of biochemistry is an exciting challenge. First, and of prime importance, the material in an introductory biochemistry text must be current. Discoveries and developments are taking place constantly, and even explanations change so rapidly that a determined effort must be made to keep abreast of the field. The revision process on this text was initiated only after countless hours of studying recent biochemical developments in research journals, reviews, and symposia. In an attempt to embrace current pedagogical practices, such as a considerably changed order of presentation of the subject, and to include recent developments, an extensive rewriting of the text was necessary.

The chapter order is no longer dictated by the organic nature of biochemical compounds. The material starts with the biochemistry of the cell and its components. The proteins, nucleic acids, and enzymes of the cell are then considered, prior to a study of the carbohydrates and lipids. Changes from the first edition include new chapters on vitamins and coenzymes, an introduction to metabolism, and the biochemistry of genetics. The material in each chapter has been rewritten or revised and brought up to date, consistent with current biochemical thinking. A list of suggested readings for each topic and a glossary have been added to aid the student.

In addition, several learning aids have been added to each chapter. The material is preceded by a series of objectives pointing out essential information to be learned, and the chapter is followed by a list of important terms and concepts plus selected questions and suggested readings. To assist the student in keeping abreast of current developments in the subject, each chapter includes a topic of current interest in the area of biochemistry being discussed. It is hoped that this type of presentation will stimulate students to learn and to appreciate the role of biochemistry in the health sciences.

The knowledge gained from this text should be useful to any student planning a career in medical technology, nuclear medical technology, nursing, pharmacy, dentistry, or various aspects of medicine. Also, community colleges, junior colleges, and liberal arts colleges are presently stressing a comprehensive curriculum, which often includes fundamental knowledge of biochemistry. Students majoring in related fields of chemistry and biology discover the close relation between their area of study and biochemistry. Liberal arts students find that an understanding of biochemistry is valuable in broadening their total educational perspective. Perhaps of even greater importance is the understanding of new developments in science and the rapid changes in drug therapy, genetic applications, and medical progress made possible by the knowledge of biochemistry.

A satisfactory revision of a textbook obviously depends on many individuals. Most important are the students who used the first edition and their instructors. The valuable comments, criticisms, and suggestions received from this group are greatly appreciated, and resulted in major changes in the text. More specifically, the role of co-authors Donald Burton and Darrell Eyman cannot be underestimated. Since the first edition of this text appeared, the two larger books, *Essentials of General, Organic and Biochemistry* and *A Brief Introduction to General, Organic and Biochemistry,* have both been revised by the three authors. The present text is a part of the *Essentials* book and owes much to the cooperative effort of the two co-authors. Behind every book is a valuable secretary and unpublicized co-author—in this case, Lynn Stoll, whose excellent contributions are cheerfully acknowledged and appreciated. Finally, the author is most grateful to his publisher for the opportunity to prepare this second edition.

CONTENTS

v

CONTENTS

INTRODUCTION

Scientific discoveries and advances occur so rapidly in the 1970's that they are often reported in newspapers and news magazines before they have been confirmed by other scientists. This is especially true with progress in biochemistry. Many people are interested, for example, in the most recent findings concerning dietary cholesterol and its relationship to atherosclerosis and heart disease. Triglycerides, prostaglandins, and free fatty acids are becoming common terms as people wage a continuous battle against obesity and attempt to gain more knowledge about the control exerted on metabolic processes by their body hormones. In addition to the popular concern about the action of common and readily available drugs on the biochemistry of the body, a large body of knowledge is being gathered concerning the effect on biochemical processes of pesticides, herbicides, and a multitude of chemical pollutants, which are present in the soil, our food, drinking water, lakes, and rivers. Clinical biochemistry, which emphasizes the medical aspects of the subject, has grown rapidly in recent years. This field utilizes the determination of hormones, vitamins, trace elements, enzymes and isoenzymes, and body constituents such as glucose, cholesterol, and uric acid to assist the physician in diagnosis and treatment. In their recent studies on DNA, genes, chromosomes, and the genetic code, biochemists have recognized many hereditary defects known as inborn errors of metabolism. They are recognizing the genetic implications of several enzymes in key metabolic processes and the vital importance of the mechanism of enzyme synthesis within the cell. Experiments on the genetic alteration of microorganisms by the introduction of foreign RNA or DNA molecules have created the possibility of successful genetic engineering. The future control of such experiments will be essential to our well-being. The recognition by biochemists of the controlling influence exerted on metabolism by enzymes, vitamins, coenzymes, and hormones has resulted in increased research on the effect of these agents. In addition to their individual actions, these compounds act together in so-called positive and negative feedback systems, which serve as accelerators and braking systems for multiple biochemical reactions.

THE DEVELOPMENT OF BIOCHEMISTRY

Biochemistry, as such, is a relatively young science with an interesting history. Early in the sixteenth century, Paracelsus, the son of a Swiss physician, probably carried out biochemical experimentation in his development of a new era of medical chemistry. He believed that life processes are essentially chemical in nature and that disease could be cured by the administration of chemical medicine. Lavoisier, in the latter half of the eighteenth century, recognized that the process of oxidation of foodstuff in the body required oxygen and that carbon dioxide and heat energy are evolved. His studies led to more sophisticated research on animal and human calorimetry in Germany and the United States during the nineteenth century.

From a chemical standpoint, biochemistry was an offshoot of organic chemistry and was definitely affected by historical developments in that field. Prior to the synthesis of the organic compound urea from the inorganic compound ammonium cyanate by Wöhler in 1828, the majority of chemists were convinced that all organic compounds originated in living matter. A mysterious vital force was thought necessary for the production of these compounds by the living cell. Once the vital force theory was overthrown, organic chemistry and, with it, biochemistry developed in a satisfactory manner.

Pasteur (1822–1895) dedicated his life to a study of microorganisms and their properties. He discovered that they were living organisms that were capable of producing disease and were responsible for the fermentation of sugar to produce alcohol. The yeast used in the wine industry was thought to contain a vital principle essential to the process. The Büchner brothers in 1897 ground the yeast with sand to disrupt all the living cells and were able to ferment sugar into alcohol with an extract from the broken cells. These experiments not only disproved the vital force theory in living cells but served as a basis for the modern concept of enzymes in the cell.

Biochemistry, from a medical standpoint, had its beginnings in physiology. At the end of the nineteenth and beginning of the twentieth century, a few outstanding physiology laboratories in France, England, Germany, and the United States were carrying out research in chemical physiology or physiological chemistry. With the organization of the American Society of Biological Chemists in 1906, biochemistry became a separate and recognized field of chemistry.

Since 1960 and into the 1970's progress in biochemistry has been only a little less than phenomenal. This was due in great measure to the large sums of money made available by the government for fundamental research. This support by the government resulted in an increase of training programs and an increased production of biochemical specialists.

Many biochemists directed their efforts toward a clarification of the metabolic cycles that occur in the cells and tissues. Enzymes, coenzymes, and intermediate metabolites in the various pathways were identified. In addition, the role of adenosine triphosphate, ATP, in the energy relationships of the cycles was established. The importance of the Krebs, or tricarboxylic acid cycle, was pointed out forcefully when it was shown that amino acids and fatty acids are oxidized in the body to produce intermediate compounds identical with those in the Krebs cycle. This information provided a bridge between carbohydrate, lipid, and protein metabolism and a common mechanism for the liberation of energy from all foodstuffs.

With the aid of the electron microscope the biochemists have focused their attention on the components of the individual cells and the reactions that occur in these components. As a prelude to a study of cellular reactions, the structure of cellular proteins and nucleic acids was investigated. The primary polypeptide structure of proteins had been proposed years earlier by Emil Fischer, a German biochemist. Sanger in England reinvestigated protein structure and in a series of painstaking research studies established the complete amino acid sequence of the protein hormone insulin. Other protein molecules were investigated to determine their amino acid sequence and details of their structure. Pauling and Corey proposed a helical arrangement of amino acids in the secondary structure of protein molecules and this stimulated Watson and Crick to apply the idea to the structure of nucleic acids such as DNA and RNA.

Synthesis of protein molecules within the cell is a problem that has recently involved untold hours of biochemical research. The mechanism involves various types of RNA molecules and chromosomal DNA molecules. The mystery of the process by which the cell synthesizes complete protein molecules such as insulin from amino acids has led to a study of biochemical genetics. The genetic role of DNA as a bearer of the genes of chromosomes and a controlling influence in the synthesis of proteins has been clarified. In addition, the relationship between DNA in the nucleus of cells and RNA in the cytoplasm and ribosomes has been studied extensively. As a result of this subcellular research, the location and the nature of biochemical reactions and metabolic pathways within the subcellular components continue to be subjects of prime interest to the modern biochemist.

Although the first scientific journal devoted to biochemistry was published in Germany in 1879, three new journals, the *Journal of Biological Chemistry* (United States), the *Biochemical Journal* (England), and the *Biochemische Zeitschrift* (Germany), were first published in 1906. From 1906 until the middle 1940's biochemistry was concerned with the study of the whole body, whether human or animal, and research was done on nutrition, vitamins, amino acids, fatty acids, and the chemistry of physiological function. Widespread nutritional research was carried out on experimental animals including mice, rats, hamsters, guinea pigs, and rabbits. Amino acids that were essential for growth in animals, essential fatty acids and the chemistry of vitamins were all established during this period. In the 1940's and 1950's animals and their organs gave way to thin tissue slices of liver, kidney, and brain, and reactions were carried out in Warburg respiration flasks measuring oxygen uptake and carbon dioxide production of respiring or expiring tissues. The experimental techniques that were applied in the study of tissue slices permitted a more careful control of the chemical compounds that were added to the system with the result that chemical mechanisms in the tissues could be investigated. This era marked the development of the early metabolic cycles proposed by Embden and Meyerhof, Krebs, and Krebs and Hensleit.

Chapter 1

BIOCHEMISTRY OF THE CELL

The *objectives* of this chapter are to enable the student to:

1. Recognize that the definition of the living cell as proposed by Lehninger forms the basis for the study of biochemistry.
2. Differentiate between anabolism and catabolism and understand their relation to metabolism.
3. Describe the nature of the major subcellular components.
4. Recognize the similarities and differences between plant and animal cells.
5. Describe the method of separation of cellular components.
6. Outline with a diagram the functions of the major subcellular components.

The ultimate aim of biochemistry is a clear understanding of the chemical reactions that occur in the living cell and their relation to cellular function and structure. The use of radioisotope-labeled compounds and cytochemical techniques has assisted the biochemist in locating the cellular sites of specific reactions, especially those involving enzymes. The **light microscope** has been invaluable in the study of staining reactions and the rough morphology of the cell. With the advent of the **electron microscope** the fine structure of the cell was revealed, and an entirely new area of biochemical research was made available. A more complete understanding of the relationship of structure to function is now within the grasp of biochemists. In fact, Lehninger recently defined the living cell as a self-assembling, self-adjusting, self-perpetuating isothermal system of molecules that exchanges matter and energy with its environment. This system carries out many consecutive organic reactions that are promoted by enzymes produced by the cell. It operates on the principle of maximum economy of parts and processes, and its precise self-replication is ensured by a linear molecular code. The study of biochemistry concerns the many facets of Lehninger's definition, which should be referred to as each chapter of the section is studied.

The macromolecules, which make up the components of the cell, are formed from low molecular weight precursors that are transformed into the unit structures of biochemistry from metabolic intermediates in the living cell. The unit structures are organic molecules that serve as building blocks for the more complex macromolecules of the cell, which include proteins, carbohydrates, lipids, and nucleic acids. These macromolecules are further assembled into cellular systems such as enzyme complexes and ribosomes. The final

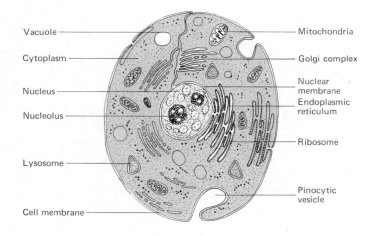

assembly results in subcellular components or organelles such as the nucleus and mitochondria. Two major types of cells based on size and complexity are the prokaryotic and eukaryotic cells. The **prokaryotes** are very small, simple cells having only a single membrane without nucleus or organelles. **Eukaryotes** are larger and more complex, 1000 to 10,000 times the cell volume of a prokaryote. They are present in all higher plant and animal organisms. In addition to the cell membrane, they contain a membrane-surrounded nucleus and organelles such as the mitochondria, endoplasmic reticulum, and Golgi bodies. An example of a prokaryote is the bacterium, *E. coli*, whereas a rat liver cell or hepatocyte would represent a eukaryote.

The chemical activities of the intact cell which provide for its growth, maintenance, and repair constitute the process of **metabolism.** Cells undergo constant change by taking in new substances, altering them chemically, building new cellular materials, and transforming the potential energy present in the molecules of proteins, carbohydrates, and fats into chemical and kinetic energy and heat. Metabolic processes are commonly divided into anabolic and catabolic processes. **Anabolism** refers to processes in which simpler substances are combined to form more complex substances, resulting in the storage of energy, the production of new cellular material, and growth. **Catabolism** is the process of breaking down these complex substances with the release of energy and consumption of cellular materials.

Cells differ in size, appearance, and structure, depending on their function, but a typical animal cell has the features illustrated in Figure 1–1. A typical plant cell (Fig. 1–2) also includes the same structures. To understand the function of a plant, animal, or bacterial cell we must first study the subcellular components. Structures common to all

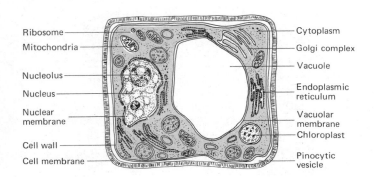

Figure 1-2 A diagram of a typical plant cell.

cells include the cell membrane, nucleus, mitochondria, endoplasmic reticulum, ribosomes, and Golgi apparatus. Fortunately for cytological studies, the electron microscope can be focused on each subcellular component to reveal its structural details. A brief description of the biochemical processes occurring in each component may refer to compounds of which the structures are not well known to the student, but which will be described in later chapters.

SUBCELLULAR COMPONENTS

Cell Membrane

All the subcellular components of a cell are contained within a definite cell wall or membrane. This membrane plays a vital role in the passage of nutrient and waste material into and out of the cell. In addition to the cell membrane, plant cells have rigid walls that surround and protect the membrane. The cell walls consist of cellulose and other polysaccharides. The **cell membrane,** also called the **plasma membrane,** behaves as though it has ultramicroscopic pores, and the size of the pores determines the maximum size of the molecules that can pass through the membrane. Factors such as electric charge, the lipid solubility of the material, and the number of water molecules bound to the particle also determine whether the substance will pass through the membrane. The plasma membranes of animal, plant, and bacterial cells appear to be constructed of protein and phospholipids. A middle layer of phospholipid molecules 6.0 nm thick is surrounded by an outer and inner layer of protein molecules 3.0 nm thick. This pattern of protein-phospholipid-protein in membranes is called the unit membrane and appears to be the generally accepted structure of the cell or plasma membrane.

The permeability of living membranes has never been adequately explained. Although many cells are bathed in a fluid rich in sodium and chloride ions and low in potassium ions, the cell contents are rich in potassium ions and low in sodium and chloride ions. Membranes must therefore be involved in a process of "active transport," or the movement of ions or molecules from a region of low concentration across the membrane into a region of higher concentration. This movement against a concentration gradient involves enzymes such as ATPase and energy in the form of adenosine triphosphate (ATP). The different rates of absorption of monosaccharides and amino acids from the small intestine emphasize the importance of the membrane in selective permeability toward small ions and molecules. In general, large molecules do not pass directly through the cell membrane; however, they may be taken into the cytoplasm by pinocytosis. **Pinocytosis** starts with the indentation of the cell membrane to form a pinocytic vesicle (Fig. 1–1) that engulfs the large molecules and then closes with the formation of a vacuole or lysosome that moves into the cytoplasm of the cell.

Cytoplasm

The **cytoplasm** is the general protoplasmic mass in which the definite subcellular components described above are embedded. At present all of the essential compounds and macromolecules in the cell not associated with definite particles are thought to exist in the cytoplasm. Many soluble enzymes are found in the cytoplasm, particularly those associated with the conversion of glucose to pyruvic or lactic acids. Considerable research remains to be done on the components of the cell and the cytoplasm with respect to enzyme and coenzyme distribution and their role in various metabolic reactions.

Nucleus

The nucleus is a small spherical or oval-shaped organelle, separated from the surrounding cytoplasm by a double layer of unit membrane called the nuclear membrane,

Outer membrane ———

Inner membrane ———

Cristae ———

FIGURE 1-3   Tridimensional diagram of mitochrondia showing outer and inner membranes with the cristae.

which regulates the flow of material into and out of the nucleus. In many cells the outer membrane is connected with the nuclear membrane by one or more channels through the cytoplasm. In addition there is usually a connection between the endoplasmic reticulum and the double-layered nuclear membrane. It has long been recognized that the **nucleus** serves as a site for the transmission and regulation of hereditary characteristics of the cell. This control is an essential feature of the **chromosomes** that are composed of **deoxyribose nucleic acid** (DNA), and basic protein (see p. 31). The nucleus also contains one or more small, dense, round bodies called nucleoli. These bodies contain DNA and **ribose nucleic acid** (RNA), and appear to be involved in the synthesis of RNA and proteins.

Mitochondria

These subcellular particles are shaped like an elongated oval 2 to 7μ in length and 1 to 3μ in diameter (Fig. 1–3). It is possible to stain mitochondria and observe them under a light microscope; however, the electron microscope reveals their fine structure. The mitochondria are bounded by double membranes; the outer layer forms a smooth boundary, while the inner layer is folded repeatedly into parallel ridges or plates that extend into the center of the mitochondrial cavity. These shelf-like inner folds are called **cristae** and contain the enzymes that are involved in the formation of ATP by the process of oxidative phosphorylation. The liquid within the inner compartment of the mitochondria contains protein, neutral fat, phospholipids, nucleic acids, and the enzymes of the Krebs cycle. In contrast to the nucleus, the nucleic acids in the mitochondria are mostly RNA with only small amounts of DNA. The energy production in the form of ATP from the major oxidative processes and oxidative phosphorylation is the prime function of the **mitochondria** and is the reason they are called the *powerhouses of the cell.*

TOPIC OF CURRENT INTEREST

MITOCHONDRIA—THE POWERHOUSES OF THE CELL

Mitochondria are complex cellular organelles, the main function of which is the production of useful energy in the form of the high-energy phosphate groups of ATP. All the components required for the process of oxidative phosphorylation (formation of ATP) and the Krebs cycle are contained within these subcellular particles. In the cell, mitochondria are often located near structures that require the energy of ATP or near a source of fuel for oxidation, such as the cytoplasmic fat droplets. In the flight muscles of insects, for example, the mitochondria are lined up along the contractile fibers and the ATP molecules after formation have only a short distance to diffuse to the fibers that require their energy.

The structure and function of the mitochondria of various cells has been the subject of intensive research. Rat liver cell mitochondria have been extensively studied and may serve as a model for mitochondrial structure. They are shaped like a football 2 to 4μ in length and 1 to 2μ in diameter (similar in size to the bacterium *E. coli*). Their walls consist of double membranes, an outer membrane that is smooth and bag-shaped and an inner membrane that is folded into ridges or plates that extend into the center of the mitochondrial cavity. These inner folds are called **cristae;** the cristae greatly increase the surface area of the inner membrane of the mitochondria. The gel-like fluid inside the mitochondria, called the **matrix,** is rich in protein (about 50 per cent) and also contains neutral fat, phospholipids, and nucleic acids.

The outer membrane, the inner membrane, and the matrix of the rat liver mitochondria have been analyzed for their composition and enzyme content. An example of the enzymes in the outer membrane is monoamine oxidase, which catalyzes the oxidation of monoamines such as epinephrine (adrenaline). The inner membrane contains cytochromes and all the enzymes and factors required for the process of ATP production by oxidative phosphorylation. The enzymes and coenzymes associated with the Krebs cycle are located in the matrix. The inner membrane of rat liver mitochondria is a particularly complex system containing more than 50 biologically active protein molecules associated with the phospholipid bilayer membranes. These proteins include enzymes concerned in ATP synthesis, electron-transferring enzymes and proteins, several dehydrogenase enzymes, and the protein components of transport systems for metabolites and inorganic ions.

The importance of the mitochondria will become more evident as we study their involvement in electron transport and oxidative phosphorylation and the Krebs cycle in carbohydrate metabolism. With their collection of biologically active proteins and enzymes and cofactors involved in the production of useful energy in the form of ATP, they are truly *the powerhouses of the cell.*

Chloroplasts

Plant cells contain highly pigmented particles 3 to 6μ in diameter called **chloroplasts** (Fig. 1–2). These particles contain the green pigment chlorophyll and play a major role in photosynthesis. Inside the chloroplast membrane is a series of laminated membrane structures called **grana.** Chlorophyll and lipids are concentrated in the grana which are active in the photosynthetic process. The structure and function of chloroplasts parallel those of mitochondria.

Endoplasmic Reticulum and Ribosomes

The **endoplasmic reticulum** is composed of a network of interconnected sheets of membrane-like tubules or vesicles. In some areas of the cytoplasm the membranes are covered with dark round bodies, about 0.15μ in diameter, which contain 80 per cent of the RNA in the cell and have, therefore, been termed **ribosomes.** This type is known as granular or rough endoplasmic reticulum, in contrast to the agranular or smooth endoplasmic reticulum, which does not have ribosomes adsorbed on its surface. The ribosomes are part of a heterogeneous group of smaller particles called microsomes. The specific particles, the ribosomes, may be separated from the microsomal fraction; they have as their main function the synthesis of proteins within the cell.

The Golgi Complex

This is also called the Golgi body and consists of an orderly array of flattened sacs with smooth membranes associated with small vacuoles of varying size. Although it is similar to the smooth endoplasmic reticulum, it exhibits different staining properties and its membrane has a different composition. The **Golgi complex** is often connected to the cell membrane by a channel and serves as a way station in the transport of substances produced in other subcellular particles. In liver cells, for example, the Golgi complex is usually located close to the small bile canals and is involved in the transport and excretion of substances such as bilirubin glucuronide into the bile. In other cells the complex serves as a temporary storage place for proteins produced in the endoplasmic reticulum and aids in the transport and secretion of these proteins through the cell membrane.

Lysosomes

These particles are spherical in shape with an average diameter of 0.4μ. They contain several soluble hydrolytic enzymes (hydrolases) that exhibit an optimum pH in the acid range. The lysosomal membrane is lipoprotein in nature and prevents the enzymes from escaping into the cellular cytoplasm. The membrane also prevents the substrates for the enzymes from entering the cell. When the cell is injured and the membrane is broken, the released enzymes cause cellular breakdown. In autolysis of tissue, whether normal (as involution of the thymus gland at puberty), pathological, or post-mortem, the lysosomal enzymes destroy cellular tissue. In fact, one of the main functions suggested for these particles is to help clear tissues of dead cells. The processes of phagocytosis and pinocytosis involve the engulfment of foreign material into vesicles or vacuoles and the digestion of this material. These particles may be converted into lysosomes to assist in the hydrolysis of phagocytosed material.

Vacuoles and Vesicles

These particles are roughly spherical in shape and vary in size from 0.1 to 0.7μ in diameter. They are often found close to the Golgi apparatus and to channels involved in the entrance and excretion of material to and from the cell. Vacuoles may serve as temporary storage sacs, or as bodies involved in the removal of foreign material from the cell.

SEPARATION OF CELLULAR COMPONENTS

Since in most instances it is impossible to study the reactions in a single cell or in a particular particle within that cell, several schemes for the separation of cellular

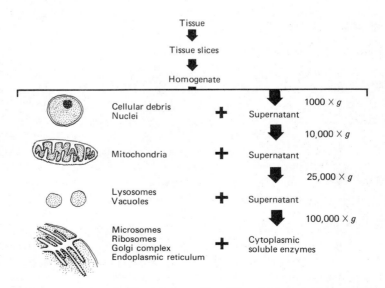

Figure 1-4 The separation of cell components by differential centrifugation. (Modified from Bennett and Frieden: Modern Topics in Biochemistry, The Macmillan Co., 1966.)

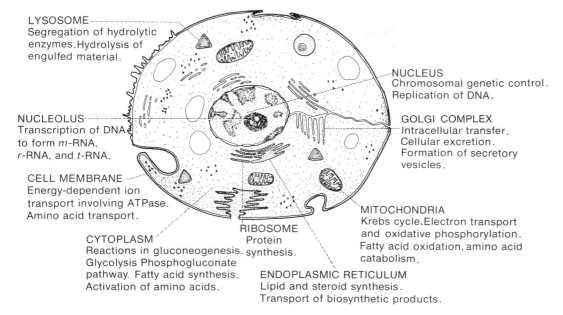

LYSOSOME
Segregation of hydrolytic enzymes. Hydrolysis of engulfed material.

NUCLEUS
Chromosomal genetic control. Replication of DNA.

NUCLEOLUS
Transcription of DNA to form m-RNA, r-RNA, and t-RNA.

GOLGI COMPLEX
Intracellular transfer. Cellular excretion. Formation of secretory vesicles.

CELL MEMBRANE
Energy-dependent ion transport involving ATPase. Amino acid transport.

MITOCHONDRIA
Krebs cycle. Electron transport and oxidative phosphorylation. Fatty acid oxidation, amino acid catabolism.

CYTOPLASM
Reactions in gluconeogenesis. Glycolysis Phosphogluconate pathway. Fatty acid synthesis. Activation of amino acids.

RIBOSOME
Protein synthesis.

ENDOPLASMIC RETICULUM
Lipid and steroid synthesis. Transport of biosynthetic products.

FIGURE 1-5   Biochemical functions of cellular components.

components have been proposed. To study enzyme systems in the mitochondria, for instance, it is desirable to have a large number of functioning mitochondria free from cytoplasm and other cell particles. The isolation procedure begins with thin slices of the proper tissue or organ which are homogenized by grinding the tissue in a glass or Teflon homogenizer in cold isotonic sucrose solution. The homogenate, which consists mostly of disrupted cells, is then subjected to **differential centrifugation** to separate the subcellular particles. An example of a general scheme for the separation of the essential components of a cell is shown in the accompanying diagram (Fig. 1–4). The gravitational force of centrifugation in the scheme is represented as g (for example, $10,000 \times g$); the time of centrifugation is approximately 10 to 15 minutes for all steps but the final centrifugation, which requires one hour.

BIOCHEMICAL FUNCTION OF CELLULAR COMPONENTS

Many investigators have concentrated their research activities on the biochemical reactions that occur in a specific subcellular particle. The mitochondria and ribosomes particularly have been the subject of several research studies. It is obviously not possible at present to reconstruct the exact biochemical functions of the intact cell, but a combination of cytochemical techniques and research on reactions within the separated particles provide a greater understanding of the overall process. Some of the biochemical functions that have been associated with cellular components are outlined in Figure 1–5.

IMPORTANT TERMS AND CONCEPTS

anabolism
catabolism
chloroplasts
cytoplasm
differential centrifugation
electron microscope
Golgi complex

lysosomes
membrane
metabolism
mitochondria
nucleus
vacuole

QUESTIONS

1. What major instrumental development enabled the biochemist to study subcellular components?

2. Give a definition of the living cell.

3. Distinguish between prokaryotes and eukaryotes.

4. Explain the relation of anabolism and catabolism to metabolism.

5. What is the nature of the cell membrane? What evidence is there for selective permeability of the cell membrane?

6. The membrane surrounding the nucleus not only has pores but has a direct connection to the cytoplasm through the endoplasmic reticulum. Is this arrangement an advantage to the cell? Why?

7. Biochemists have intensively studied the mitochondria for several years. Why should they be so interested in this subcellular particle?

8. In an electron micrograph of the cell, how can rough endoplasmic reticulum be differentiated from smooth? What is one of the major functions of the rough form?

9. Discuss the nature and function of the Golgi apparatus.

10. What is the nature and function of the lysosomes?

Chapter 2

PROTEINS

The *objectives* of this chapter are to enable the student to:

1. Differentiate between proteins, carbohydrates, and fats on the basis of their elementary composition.
2. Explain why the structures of amino acids are often written in the ionized form.
3. Explain why naturally occurring amino acids are called alpha amino acids.
4. Recognize a sulfur-containing, an aromatic, and a heterocyclic amino acid.
5. Graphically illustrate the pK_1, the pK_2, and the isoelectric point of an amino acid like glycine.
6. Illustrate the formation and hydrolysis of a dipeptide.
7. Describe the separation of amino acids by ion exchange chromatography.
8. Describe the analysis of the insulin molecule by Sanger.
9. Distinguish between the primary, secondary, and tertiary structures of a protein.
10. Illustrate the importance of hydrogen bonding in the secondary structure of proteins.
11. Describe how to determine the amount of protein in a solution.
12. Describe three methods of precipitating a protein out of a solution.
13. Write the equation for the precipitation of a protein by a heavy metal.

The chemistry of living organisms is mainly concerned with the **proteins** of their cells. Half of the solid matter of cells consists of protein, and it has been estimated that the average cell contains about 3000 different kinds of proteins. These proteins are involved in anabolism and catabolism in the cell, the synthesis of body tissues, enzymes, certain hormones, and protein components of the blood. They also function in transportation systems in the tissues, body motion, and protection from bacterial invasion.

Proteins are made by plant cells by a process starting with photosynthesis from carbon dioxide, water, nitrates, sulfates and phosphates. The complicated process of synthesis has not as yet been completely elucidated. Animals can synthesize only a limited amount of protein from inorganic sources and are mainly dependent on plants or other animals for their source of dietary protein. Proteins are used in the body for growth of new tissue, for maintenance of existing tissue, and as a source of energy. When used for energy, they are broken down by oxidation to form simple substances such as water, carbon dioxide, sulfates, phosphates, and simple nitrogen compounds that are excreted from the body. These same products are formed in decaying plant and animal matter. The simple nitrogen compounds such as amino acids and urea are converted into ammonia,

9

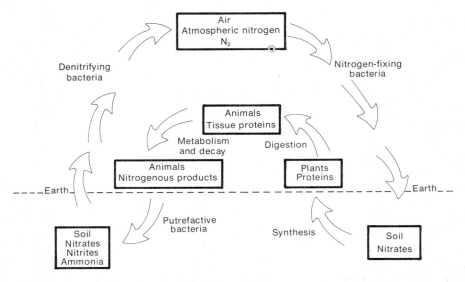

FIGURE 2-1   A simple diagram showing the events occurring in the nitrogen cycle.

nitrites, and nitrates. The growing plants then use these inorganic compounds to form new proteins, and the cycle is completed (Fig. 2–1).

ELEMENTARY COMPOSITION

The five elements that are present in most naturally occurring proteins are **carbon, hydrogen, oxygen, nitrogen,** and **sulfur.** There is a wide variation in the amount of sulfur in proteins. Gelatin, for example, contains about 0.2 per cent, in contrast to 3.4 per cent in insulin.

Other elements, such as phosphorus, iodine, and iron, may be essential constituents of certain specialized proteins. Casein, the main protein of milk, contains phosphorus, an element of utmost importance in the diet of infants and children. Iodine is a basic constituent of the protein in the thyroid gland and is present in sponges and coral. Hemoglobin of the blood, which is necessary for the process of respiration, is an iron-containing protein. Most proteins show little variation in their elementary composition; the average content of the five main elements is as follows:

Element	Average Per Cent
Carbon	53
Hydrogen	7
Oxygen	23
Nitrogen	16
Sulfur	1

The relatively high content of nitrogen differentiates proteins from fats and carbohydrates.

MOLECULAR WEIGHT

Protein molecules are very large, as indicated by the approximate formula for oxyhemoglobin:

$$C_{2932}H_{4724}N_{828}S_8Fe_4O_{840}$$

The molecular weight of oxyhemoglobin would thus be about 68,000. The com.
protein egg albumin has a molecular weight of about 34,500. In general, protein molecu
have weights that vary from 12,000 to 50,000,000. Their extremely large size can read
be appreciated when they are compared with the molecular weight of a fat such as
tripalmitin, which is 807, of glucose, which is 180, or of inorganic ions such as sodium
and chloride (Fig. 2–2).

AMINO ACIDS

In addition to their large size, protein molecules are also very complicated. Like
any complex molecule, they may be broken down by hydrolysis into smaller molecules
whose structure is more easily determined. As will be illustrated later, the hydrolysis
involves the simple splitting of an amide group. Common reagents used for the hydrolysis
of proteins are acids (HCl and H_2SO_4), bases (NaOH), and enzymes (proteases). The simple
molecules that are formed by the complete hydrolysis of a protein are called **amino acids.**

Before considering the properties and reactions of proteins, it may be well to study
the individual amino acids. An amino acid is essentially an organic acid that contains
an amino group. If a hydrogen is replaced by an amino group on the carbon atom that
is next to the carboxyl group in acetic acid, CH_3COOH, the simple amino acid **glycine**
will be formed.

$$\underset{\text{Undissociated form}}{\overset{\displaystyle CH_2COOH}{\underset{\displaystyle NH_2}{|}}} \qquad \underset{\text{Ionized form}}{\overset{\displaystyle CH_2COO^-}{\underset{\displaystyle NH_3{}^+}{|}}}$$

Amino acids are sometimes represented in the undissociated form to emphasize the
amino and carboxyl groups; however, in body fluids the ionized forms predominate around
a pH of 7.0. In the undissociated form, the carbon atom next to the carboxyl group is called
the alpha (α) carbon, and since all amino acids have an amino group attached to the alpha
carbon atom, they are known as **alpha amino acids.**

FIGURE 2-2   Relative dimensions of various
protein molecules. (After Oncley, J. L.: Confer-
ence on the Preservation of the Cellular and
Protein Components of Blood, published by the
American Red Cross, Washington, D. C.)

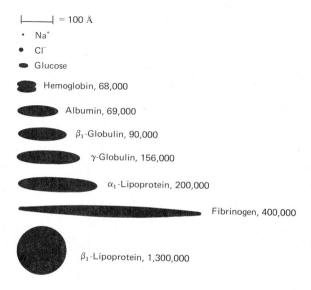

The amino acids are usually divided into groups according to their chemical structures. In the following tabulation the common name for each amino acid is followed by the abbreviation used in sequence and structure models. The ionized forms existing at pH 6.0 to 7.0 are presented and the unit common to all amino acids, consisting of the carboxyl group, alpha carbon atom, and alpha amino group, is emphasized.

Aliphatic

Glycine (Gly)

$$H-\underset{\underset{NH_3^+}{|}}{C}HCOO^-$$

Alanine (Ala)

$$CH_3-\underset{\underset{NH_3^+}{|}}{C}HCOO^-$$

Valine (Val)

$$CH_3-\underset{\underset{CH_3}{|}}{C}H-\underset{\underset{NH_3^+}{|}}{C}HCOO^-$$

Leucine (Leu)

$$CH_3-\underset{\underset{CH_3}{|}}{C}H-CH_2-\underset{\underset{NH_3^+}{|}}{C}HCOO^-$$

Isoleucine (Ile)

$$CH_3-CH_2-\underset{\underset{CH_3}{|}}{C}H-\underset{\underset{NH_3^+}{|}}{C}HCOO^-$$

Serine (Ser)

$$\underset{\underset{OH}{|}}{C}H_2-\underset{\underset{NH_3^+}{|}}{C}HCOO^-$$

Threonine (Thr)

$$CH_3-\underset{\underset{OH}{|}}{C}H-\underset{\underset{NH_3^+}{|}}{C}HCOO^-$$

Basic

Lysine (Lys)

$$\underset{\underset{NH_2}{|}}{C}H_2-CH_2-CH_2-CH_2-\underset{\underset{NH_3^+}{|}}{C}HCOO^-$$

Arginine (Arg)

$$\underset{\underset{NH}{\|}}{\overset{\overset{NH_2}{|}}{C}}-NH-CH_2-CH_2-CH_2-\underset{\underset{NH_3^+}{|}}{C}HCOO^-$$

Acidic

Aspartic acid (Asp)

$$^-OOC-CH_2-\underset{\underset{NH_3^+}{|}}{C}HCOO^-$$

Glutamic acid (Glu)

$$^-OOC-CH_2-CH_2-\underset{\underset{NH_3^+}{|}}{C}HCOO^-$$

Acidic amides

Asparagine (Asn)

$$NH_2OC-CH_2-\underset{\underset{NH_3^+}{|}}{C}HCOO^-$$

Glutamine (Glu)

$$NH_2OC-CH_2-CH_2-\underset{\underset{NH_3^+}{|}}{C}HCOO^-$$

Sulfur-containing

Cysteine (Cys)

$$HS-CH_2-\underset{\underset{NH_3^+}{|}}{C}HCOO^-$$

Cystine (Cys-Cys)

$$S-CH_2-\underset{\underset{NH_3^+}{|}}{C}HCOO^-$$
$$S-CH_2-\underset{\underset{NH_3^+}{|}}{C}HCOO^-$$

Methionine (Met)

$$CH_3-S-CH_2-CH_2-\underset{\underset{NH_3^+}{|}}{C}HCOO^-$$

Heterocyclic

Tryptophan (Trp)

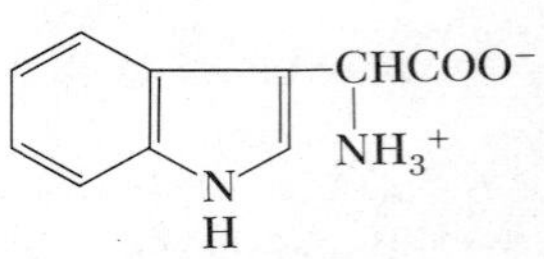

Histidine (His)

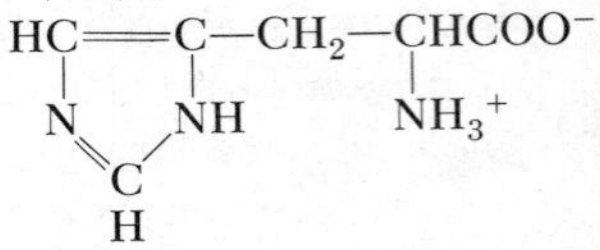

Proline (Pro)

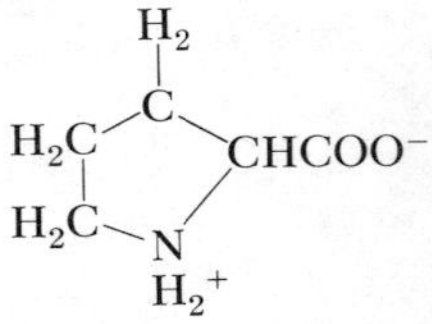

Hydroxyproline (Hypro)

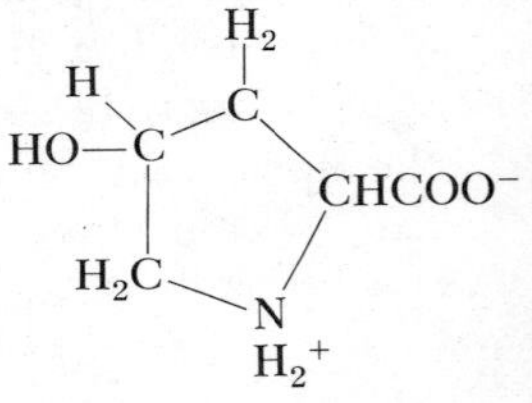

Aromatic

Phenylalanine (Phe)

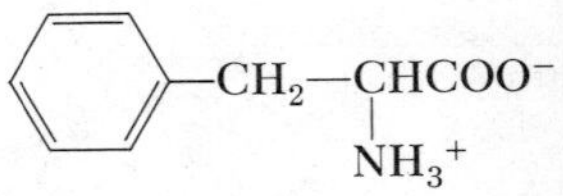

Tyrosine (Tyr)

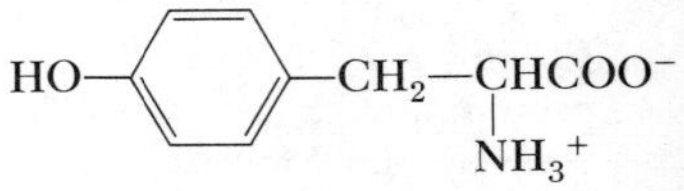

Optical Activity of Amino Acids

All amino acids except glycine contain an asymmetric carbon atom in their formulas. For this reason they may exist in the D or L form. (For an explanation of the D and L forms see p. 58.) Using alanine as an example of a simple amino acid, we may compare the D and L forms to those of glyceraldehyde and lactic acid:

$$
\begin{array}{ccc}
& H & H & O \\
& C{=}O & C{=}O & C{-}OH \\
HO{-}C{\rightarrow}H & H{-}C{\rightarrow}OH & HO{-}C{\rightarrow}H \\
CH_2OH & CH_2OH & CH_3 \\
\text{L-Glyceraldehyde} & \text{D-Glyceraldehyde} & \text{L-Lactic acid}
\end{array}
$$

$$
\begin{array}{ccc}
O & O & O \\
C{-}OH & C{-}OH & C{-}OH \\
H{-}C{\rightarrow}OH & H_2N{-}C{\rightarrow}H & H{-}C{\rightarrow}NH_2 \\
CH_3 & CH_3 & CH_3 \\
\text{D-Lactic acid} & \text{L-Alanine} & \text{D-Alanine}
\end{array}
$$

Naturally occurring amino acids from plant and animal sources have the L configuration and would be designated $L(+)$ or $L(-)$, depending on their rotation of plane polarized light. D-alanine and D-glutamic acid have been obtained from microorganisms, especially from their cell walls.

Amphoteric Properties of Amino Acids

Amino acids behave both as weak acids and as weak bases, since they contain at least one carboxyl and one amino group. Substances that ionize as both acids and bases in aqueous solution are called **amphoteric.** An example would be glycine, in which both the acidic and basic groups are ionized in solution to form dipolar ions or **zwitterions.**

$$
\begin{array}{c}
CH_2COO^- \\
| \\
NH_3^+
\end{array}
$$

The glycine molecule is electrically neutral, since it contains an equal number of positive and negative ions. The zwitterion form of glycine would therefore be isoelectric, and the pH at which the zwitterion does not migrate in an electric field is called the **isoelectric point.** Amphoteric compounds will react with either acids or bases to form salts. This is best illustrated by use of the zwitterion form of the amino acid.

$$
CH_3{-}\underset{\underset{NH_3^+}{|}}{CH}{-}COO^- + HCl \rightarrow CH_3{-}\underset{\underset{NH_3^+Cl^-}{|}}{CH}{-}COOH
$$

$$
CH_3{-}\underset{\underset{NH_3^+}{|}}{CH}{-}COO^- + NaOH \rightarrow CH_3{-}\underset{\underset{NH_2}{|}}{CH}{-}COO^- \ Na^+ + H_2O
$$

From these equations it can be seen that the addition of a H^+ to an isoelectric molecule results in an increased positive charge (NH_3^+), since the acid represses the ionization of the carboxyl group. Conversely, the addition of a base to an isoelectric molecule results in an increased negative charge (COO^-), since the base represses the ionization of the amino group. Since proteins are composed of amino acids, they are amphoteric substances with specific isoelectric points and are able to neutralize both acids and bases. This property of proteins is responsible for their buffering action in blood and other fluids.

Titration of Amino Acids

Since amino acids serve as buffers in blood and other body fluids, their dipolar ionic nature should result in at least two dissociation constants when they react with acid or base. The **Henderson-Hasselbalch equation** for a simple buffer represents the dissociation constant, or pK, as the pH at which equal concentrations of the salt and acid forms of the buffer exist in solution.

$$pH = pK + \log \frac{[salt]}{[acid]} = pK + \log \frac{1}{1} = pK$$

The simple amino acid glycine may be used as an example of amino acids or proteins as buffers. When glycine in solution is titrated with acid or base, the molecule changes from the zwitterion form to a form dissociating as either a charged amino or carboxyl group.

$$\underset{\substack{| \\ NH_3^+ \\ \text{Acid solution}}}{CH_2-COOH} \underset{pK_1}{\overset{K_1}{\rightleftarrows}} \underset{pH=2.4}{\overset{+}{H}} + \underset{\substack{| \\ NH_3^+ \\ \text{Zwitterion} \\ \text{Isoelectric point} \\ pH=6.0}}{CH_2-COO^-} + OH^- \underset{pK_2}{\overset{K_2}{\rightleftarrows}} \underset{\substack{| \\ NH_2 \\ \text{Basic solution}}}{CH_2COO^-} \quad pH=9.6$$

A mixture of the zwitterion and glycine in acid solution would compose a buffer whose pK_1 would equal the pH of the solution when equal quantities of the two forms were present in the solution. This point pK_1 at a pH of 2.4 also corresponds to the point of half neutralization of the carboxyl group. When base is added to glycine, the pK_2, or point of half neutralization of the amino group, corresponds to a pH of 9.6. The titration of glycine with acid or base shown in Figure 2–3 emphasizes the two buffering

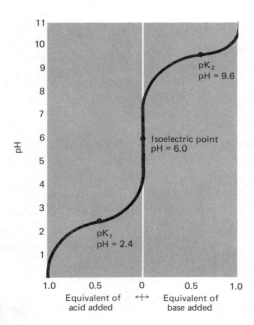

FIGURE 2-3   Titration curve for glycine.

regions and the pK values that correspond to the buffer in acid solution or in alkaline solution.

The monoamino monocarboxylic acids each exhibit two pK values and act as buffers in two pH regions. The pH of the isoelectric point can be calculated by dividing the sum of the two pK values by 2 (glycine: $2.4 + 9.6 = 12.0 \div 2 = 6.0$). More complex amino acids such as aspartic acid and lysine have three pK values and can exist in four ionized forms. The ionization of aspartic acid may be represented as follows:

$$
\begin{array}{ccccccc}
\text{COOH} & & \text{COO}^- & & \text{COO}^- & & \text{COO}^- \\
| & & | & & | & & | \\
\text{HC—NH}_3{}^+ & \underset{K_1}{\rightleftharpoons} & \text{HC—NH}_3{}^+ & \underset{K_2}{\rightleftharpoons} & \text{HC—NH}_3{}^+ & \underset{K_3}{\rightleftharpoons} & \text{HC—NH}_2 \\
| & & | & & | & & | \\
\text{CH}_2 & & \text{CH}_2 & & \text{CH}_2 & & \text{CH}_2 \\
| & & | & & | & & | \\
\text{COOH} & & \text{COOH} & & \text{COO}^- & & \text{COO}^- \\
\end{array}
$$

$$\text{pK}_1 = 2.0 \qquad \text{pK}_2 = 4.0 \qquad \text{pK}_3 = 9.8$$

Reactions of Amino Acids

The fact that amino acids can ionize as both weak acids and weak bases and contain amino groups and carboxyl groups suggests a very reactive molecule. Many of the common reactions of organic chemistry may be applied to amino acids.

Reaction with Nitrous Acid. This is the basis of the Van Slyke method for the determination of free primary amino groups.

$$
\text{R—CH—NH}_2 + \text{NaNO}_2 + \text{HCl} \rightarrow [\text{R—CH—N}{\equiv}\text{N}]\text{Cl}^- \rightarrow \text{R—CH—OH} + \text{N}_2 + \text{NaCl} + \text{H}_2\text{O}
$$
$$
\text{COOH} \qquad\qquad\qquad \text{COOH} \qquad\qquad \text{COOH}
$$

The nitrogen gas that is liberated in the reaction is collected and its volume measured. One half of this nitrogen comes from the amino acid and is used as a measure of the free amino nitrogen.

Reaction with 1-Fluoro-2,4-dinitrobenzene (FDNB). This compound, also called **Sanger's reagent**, reacts with the free amino group of an amino acid, as would an alkyl halide, to form a yellow colored dinitrophenylamino acid, DNP-amino acid.

Dinitrophenylamino acid,
or DNP-amino acid

This reaction will be found to be very important in the determination of protein structure, since the reagent reacts with the free amino group of the terminal amino acid in a protein and thus identifies the end amino acid in the structure.

Reaction with Ninhydrin. Amino acids react with ninhydrin (triketohydrindene hydrate) to form CO_2, NH_3, and an aldehyde. The amount of CO_2 that is liberated in the reaction can be used as a quantitative measure of the amino and is specific for an amino acid or compound with a free carboxyl group adjacent to an amino group. The NH_3 that is formed in the reaction combines with a molecule of reduced and a molecule of oxidized ninhydrin and forms a blue-colored compound. This compound may be measured colorimetrically for the quantitative determination of amino acids.

The reaction is complex and may be represented by two main processes:

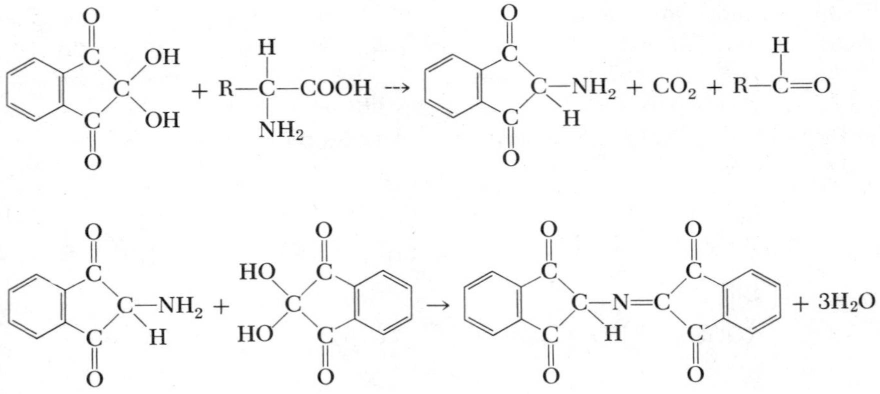

Blue-colored compound

Color Reactions of Specific Amino Acids. Certain amino acids, whether in the free form as in protein hydrolysates or combined in proteins, give specific color reactions that aid in their detection and determination. The **Millon test** depends on the formation of a red-colored mercury complex with tyrosine, whether free or in proteins, whereas tryptophan reacts with glyoxylic acid to produce a violet color in the **Hopkins-Cole test.** In the **Sakaguchi reaction,** the guanidino group in arginine forms a red color with α-naphthol and sodium hypochlorite, and cysteine and proteins that contain free sulfhydryl groups yield a red color with sodium nitroprusside. Both cystine and cysteine, free and in proteins, form a black precipitate of PbS in the **unoxidized sulfur test.**

POLYPEPTIDES

Amino acids are joined together in a polypeptide molecule by the peptide linkage. The **peptide linkage** is an amide linkage between the carboxyl group of one amino acid and the amino group of another amino acid, with the splitting out of a molecule of water. This type of linkage may be illustrated by the union of a molecule of alanine and a molecule of glycine:

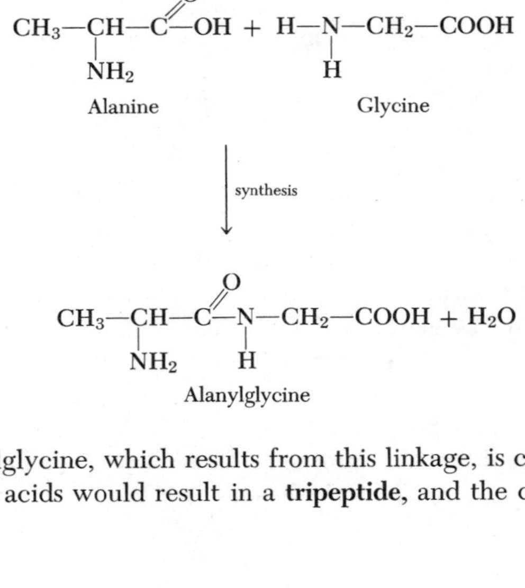

The compound alanylglycine, which results from this linkage, is called a **dipeptide.** The union of three amino acids would result in a **tripeptide,** and the combination of several amino acids by the peptide linkage would be called a **polypeptide.** Since each amino acid has lost a water molecule when it joins to two other amino acids in a polypeptide, the remaining compound is called an **amino acid residue.** Proteins may be considered as complex polypeptides. A polypeptide chain illustrating the primary structure of a protein may be represented as follows:

The R groups are the side chains of the specific amino acids in the polypeptide chain.

The Biuret Test

When a few drops of very dilute copper sulfate solution are added to a strongly alkaline solution of a peptide or protein, a violet color is produced. This is a general test for proteins and is given by peptides that contain two or more peptide linkages. **Biuret** is formed by heating urea and has a structure similar to the peptide structure of proteins:

Biuret

SEPARATION AND DETERMINATION OF AMINO ACIDS

Prior to 1950, it was thought that unraveling the combinations of the 22 different amino acids which form proteins of molecular weight from 12,000 to 50,000,000 was inconceivable. Proteins were subjected to hydrolysis and several amino acids could be separated and determined in the mixture. Amino acid sequence determination was made possible by the development of the techniques of chromatography.

Paper Chromatography

Several types of chromatographic techniques have been developed for the analysis of mixtures of molecules such as amino acids. **Paper chromatography** is relatively simple and produces excellent separation, detection, and quantitation of the individual amino acids. The technique of ascending paper chromatography is often used. When a strip of filter paper is held vertically in a closed glass cylinder with its lower end dipped in a mixture of water and an organic solvent such as butyl alcohol, phenol, or collidine, the mixture of water and organic solvent moves up the paper. If a solution containing a mixture of amino acids is added as a small spot just above the solvent level, the individual amino acids will be affected by the water phase and the organic phase as they move up the paper. A solvent partition will occur, and each amino acid will be carried to a particular location on the paper. This location depends on many factors, including the pH, the temperature, the concentration of the solvents, and the time of chromatography. After the paper is dried, it is sprayed with a solution of ninhydrin, which yields a blue to purple color with each amino acid. By comparison with known amino acids, separation of the amino acids from a hydrolysate of a protein or polypeptide fragment may be

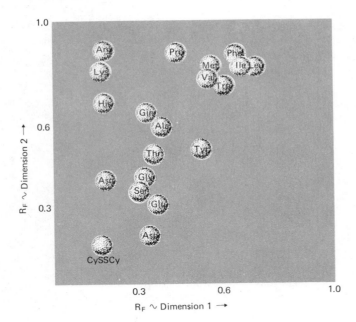

$R_F \sim$ Dimension 2 →

1.0
0.6
0.3

0.3   0.6   1.0

$R_F \sim$ Dimension 1 →

FIGURE 2-4   A schematic representation of a two-dimensional paper chromatogram. The solvent for dimension 1 is *n*-butanol-acetic acid-water (250:60:250 vol. per vol.) and for dimension 2, phenol-water-ammonia (120:20:0.3 per cent). Each solvent front moves with an R_F equal to 1 in each dimension. (After White et al.: Principles of Biochemistry, 4th ed. New York, McGraw-Hill, Inc., 1968, p. 114.)

achieved by the proper choice of conditions and solvents. This separation is often improved by a second chromatographic run during which different solvents are used and the dried paper from the first run is turned 90 degrees from the direction of the first migration. The results of a two-dimensional paper chromatography separation are shown in Figure 2-4.

A recent extension of the technique of paper chromatography involves thin-layer chromatography (see Chapter 6, p. 78). Instead of a strip of paper, the amino acids or other molecules to be separated are carried by the solvent mixture through a thin layer of cellulose powder, silica gel, and other adsorbents located on a glass plate, a plastic film, or a specially processed paper backing. The separation is much more rapid and a wide variety of reagents and dyes may be used as detection sprays. The speed of migration, sensitivity, and versatility of thin-layer chromatography make this technique a valuable addition to the tools of research in biochemistry.

Ion Exchange Chromatography

Columns of starch, cellulose powder, and alumina gels have been used to separate amino acids, but it has been difficult to isolate amino acids with similar properties from each other. More successful separations and quantitative determinations of amino acids in mixtures are achieved with **ion exchange chromatography**. Ion exchange resins are insoluble synthetic resins containing acidic or basic groups, such as —SO_3H or —OH. A sulfonated polystyrene resin may be used as a cation exchange resin by the addition of Na ions to produce —SO_3Na groups on the surface of the resin. Basic amino acids react with a cation exchange resin as follows:

$$ResinSO_3^-Na^+ + NH_3^+R \rightarrow ResinSO_3^-NH_3^+R + Na^+Cl^-$$
$$Cl^-$$

The resin is placed in a column or long glass tube, and the mixture of amino acids, which are dissolved in a small volume of buffer, is placed on top of the column. The column is washed with a buffer solution, and, as the amino acids pass down the column, the basic

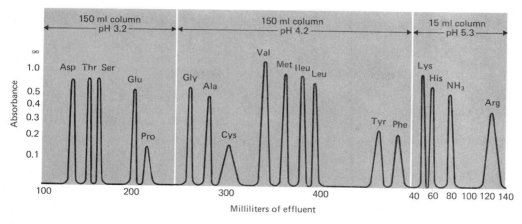

FIGURE 2-5   Chromatographic fractionation of a synthetic mixture of amino acids on columns of Amberlite IR-120. (After Moore et al.: Anal. Chem., *30*:1186, 1958.)

amino acids react with the $-SO_3^-Na^+$ groups, replacing Na^+, and are slowed in their passage. Glutamic and aspartic acids are least affected by the column and come off in the first fraction of effluent. They are followed by the neutral amino acids and finally the basic amino acids.

A more efficient removal of the different amino acids from the column is achieved by gradually increasing the pH and concentration of the eluting buffer to force the basic amino acids off the column. The elution of amino acids from their attachment to the resin by the use of sodium citrate buffer may be represented as follows:

$$ResinSO_3^-NH_3^+R + Na^+citrate^- \rightarrow ResinSO_3^-Na^+ + NH_3^+Rcitrate^-$$

By proper choice of resin, buffers, length of column, temperature, and elution rates, and collecting the eluted amino acids in fraction collectors that are coupled to automatic analyzing instruments, it is possible to obtain a quantitative amino acid analysis of a protein hydrolysate in a few hours. The **elution pattern** of representative amino acids from a cation exchange resin is illustrated in Figure 2–5.

TOPIC OF CURRENT INTEREST

THE STRUCTURE OF THE INSULIN MOLECULE ESTABLISHED BY SANGER

As stated earlier in the chapter, prior to Sanger's studies it was known that proteins were long polypeptide chains held together in different shapes by cross-links between the chains, but it was thought improbable that the exact sequence of combination of the 22 different amino acids in a specific protein molecule could be established.

The techniques available to Sanger were chromatography, enzyme hydrolysis, the FDNB method for the determination of the amino acid with the free amino group on the end of the protein chain, and the use of the enzyme carboxypolypeptidase for the determination of the amino acid with the free carboxyl group on the other end of the protein chain. In addition, the molecular weight of the unit structure of insulin was known to be about 6000.

Employing the FDNB method, Sanger first established that each insulin molecule contained two amino acids with free amino groups. He concluded, therefore, that insulin consists of two polypeptide chains. By mild oxidation he split the two chains apart and obtained evidence for three disulfide bonds in the molecule. The shorter chain (A) consisted of 21 amino acids, with glycine and asparagine as the amino acids on the N-terminal and C-terminal ends of the chain, respectively. The longer chain (B) consisted of 30 amino acids, with phenylalanine and alanine as the N-terminal and C-terminal amino acids. After a study of the oxidized polypeptide chains, his preliminary conclusion was that the molecule consisted of chains A and B held together with two disulfide links, and that chain A had an internal disulfide linkage as shown:

After carefully breaking the chains into smaller peptides by hydrolysis with enzymes or weak acid or alkali, the content and sequence of amino acids in each peptide was identified by the FDNB reaction and paper chromatography. For example, Sanger found fragments of DNP gly-ileu, DNP gly-ileu-val, and DNP gly-ileu-val-glu, which gave him a good start on the amino end of the short chain. A fragment DNP phe-val-asp-glu identified the beginning of the long chain. An example of the method used to establish the sequence of the 12 amino acids in the long chain will illustrate the tedious complexity of the problem. The following fragments were obtained by hydrolysis and identified by the FDNB method and chromatography:

Dipeptides
    ser-his leu-val glu-ala
      his-leu val-glu

Tripeptides
    ser-his-leu
      leu-val-glu
        val-glu-ala
          ala-leu-tyr
            tyr-leu-val

Higher Peptides
    ser-his-leu-val-glu
      val-glu-ala-leu   leu-val-CySO$_3$H-gly

Deduced Sequence
    -ser-his-leu-val-glu-ala-leu-tyr-leu-val-CySO$_3$H-gly-

In this fashion Sanger was able to determine the complete arrangement of amino acid residues in each chain. He proposed the sequence shown in Figure 2–6.

Amino Acid Sequence in Peptides and Proteins

After Sanger laid the foundation for the attack on the amino acid sequence of a protein molecule, other workers studied peptides and proteins. The naturally occurring

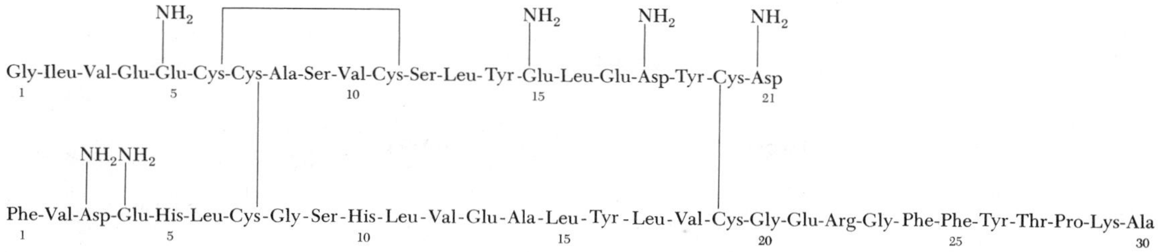

$$\text{NH}_2 \qquad\qquad \text{NH}_2 \quad \text{NH}_2 \quad \text{NH}_2$$

Gly-Ileu-Val-Glu-Glu-Cys-Cys-Ala-Ser-Val-Cys-Ser-Leu-Tyr-Glu-Leu-Glu-Asp-Tyr-Cys-Asp
1   5   10   15   21

$$\text{NH}_2\text{NH}_2$$

Phe-Val-Asp-Glu-His-Leu-Cys-Gly-Ser-His-Leu-Val-Glu-Ala-Leu-Tyr-Leu-Val-Cys-Gly-Glu-Arg-Gly-Phe-Phe-Tyr-Thr-Pro-Lys-Ala
1   5   10   15   20   25   30

FIGURE 2-6   Amino acid sequence in beef insulin.

tripeptide glutathione was shown to have a sequence as follows:

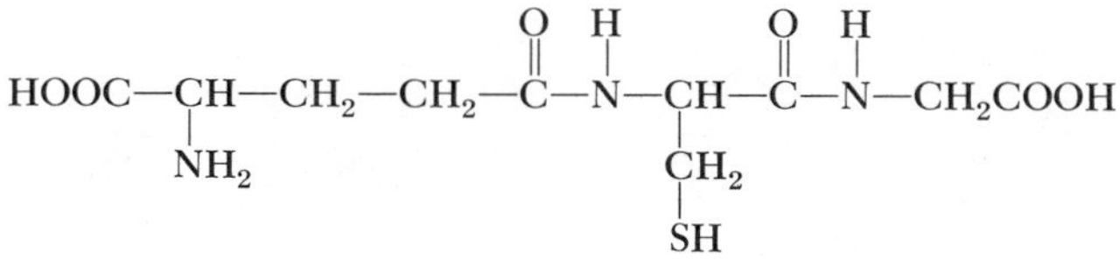

$$\text{HOOC}-\underset{\underset{\text{NH}_2}{|}}{\text{CH}}-\text{CH}_2-\text{CH}_2-\overset{\overset{\text{O}}{\|}}{\text{C}}-\overset{\overset{\text{H}}{|}}{\text{N}}-\underset{\underset{\underset{\underset{\text{SH}}{|}}{\text{CH}_2}}{|}}{\text{CH}}-\overset{\overset{\text{O}}{\|}}{\text{C}}-\overset{\overset{\text{H}}{|}}{\text{N}}-\text{CH}_2\text{COOH}$$

Glutathione (α-glutamyl-cysteinyl-glycine)

This compound in the reduced or thiol form (as shown) is essential for the normal function of the red blood cells. Vincent du Vigneaud established the exact structure and sequence in two peptide hormones, **oxytocin** and **vasopressin,** that are elaborated by the posterior lobe of the pituitary gland. Oxytocin causes contraction of smooth muscles and is used in obstetrics to initiate labor. Vasopressin constricts vessels, raising blood pressure and affecting water and electrolyte balance. Each peptide contained eight amino acids, with the disulfide bridge of cystine across four of the amino acids.

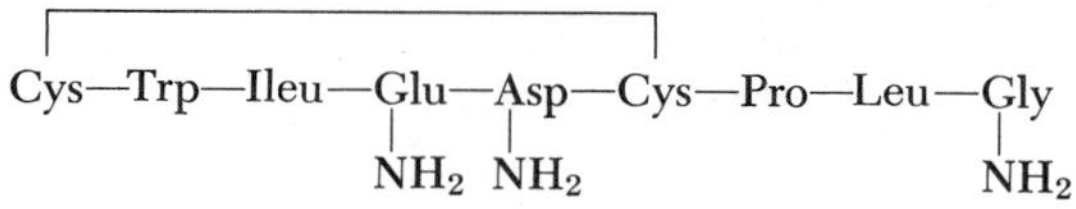

Cys—Trp—Ileu—Glu—Asp—Cys—Pro—Leu—Gly
               NH₂ NH₂   NH₂

Oxytocin

Cys—Trp—Phe—Glu—Asp—Cys—Pro—Arg—Gly
               NH₂ NH₂   NH₂

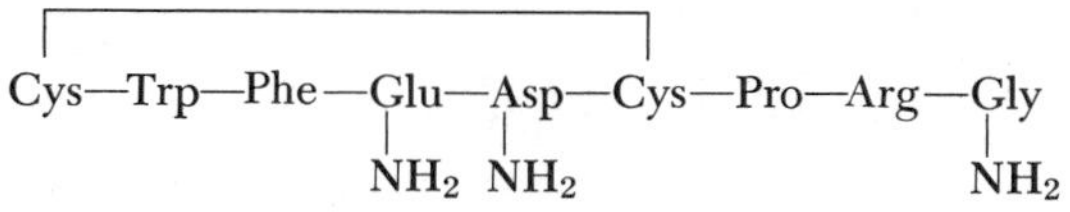

Vasopressin

The presence of two different amino acids in such a small peptide results in very different physiological activity. Du Vigneaud also succeeded in synthesizing these two molecules from amino acids and demonstrated the similar hormone activity of the synthetic peptides.

The amino acid sequence of the **adrenocorticotropic hormone, ACTH,** containing 39 amino acids, has been established. Larger protein molecules, such as the enzyme **ribonuclease** (Fig. 2–7), with 124 amino acids, the α and β polypeptide chains of **hemoglobin** (141 and 146 amino acid residues, respectively), and the **tobacco mosaic virus** protein with 158 amino acid residues, have also been characterized.

The synthesis of polypeptides such as those just described is difficult and becomes more time-consuming as the molecule increases in size. Recently, an automated solid-phase synthesis technique has been developed by Merrifield. The synthesis is carried out in a single reaction vessel by an instrument programmed to add reagents and remove

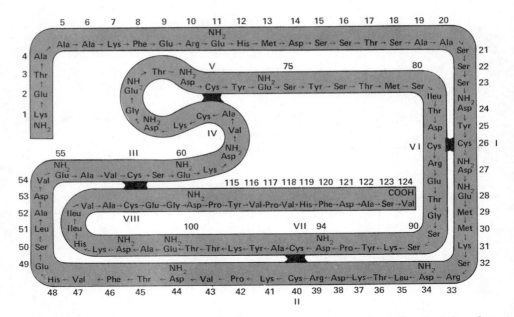

FIGURE 2-7   The complete amino acid sequence of enzyme ribonuclease. Standard three-letter abbreviations are used to indicate individual amino acid residues. (After Smyth et al.: J. Biol. Chem., 238:227, 1963.)

products at timed intervals. Briefly, the amino acid that will form the C-terminal end of the polypeptide is first attached to an insoluble resin particle. The second amino acid with its amino group blocked is added and a peptide bond formed between the NH_3 group of the first amino acid and the COOH group of the added amino acid. The blocking group is removed and the process is repeated until the desired polypeptide attached to the resin particle has been synthesized. Utilizing this technique both polypeptide chains of the insulin molecule were synthesized, and, more recently, the total synthesis of the enzyme pancreatic ribonuclease was achieved. The synthesis of such a hormone and enzyme in the laboratory may have far-reaching effects in medical therapy.

STRUCTURE OF PROTEINS

The chemical, physical, and biological properties of specific proteins depend on the structure of the molecule as it exists in the native state. Proteins range in complexity from a simple polypeptide, such as vasopressin, with biological activity, to a globular protein such as myoglobin, whose molecule includes cross linkages, helix formation, and folding and conformational forces.

PRIMARY STRUCTURE

The amino acid sequence determinations have established the exact structure of the polypeptide chain in simple proteins. The peptide linkage joining amino acids to produce a polypeptide is considered the **primary structure** of a protein (p. 16).

SECONDARY STRUCTURE

If only peptide bonds were involved in protein structure, the molecules would consist of long polypeptide chains coiled in random shapes. Most **native proteins,** however, are either fibrous or globular in nature, and consist of polypeptide chains joined together or held in definite folded shapes by hydrogen bonds. This influence of hydrogen bonding on the protein molecule is often called the **secondary structure** of the protein. Although **hydrogen bonds** may be formed between several groups on the peptide chains, the most common bonding occurs between the carbonyl and amide groups of the peptide chain backbone, as shown:

$$
\begin{array}{lll}
RCH & & HCR \\
\quad C=O\text{----}HN & \\
HN & & C=O \\
\quad HCR & RCH \\
O=C & & NH \\
\quad NH\text{----}O=C & \\
RCH & & HCR \\
\quad C=O\text{----}HN &
\end{array}
$$

Pauling (Fig. 2–8) studied the structure of the fibrous protein **α-keratin** and concluded that the polypeptide chains are regularly coiled to form a structure called the **α-helix** (Fig. 2–9A). The α-helix structure consists of a chain of amino acid units wound into a spiral which is held together by hydrogen bonds between a carbonyl group of one amino acid and the imino group of an amino acid residue further along the chain (Fig. 2–10). Each amino acid residue is 1.5 Å from the next amino acid residue, and the helix makes a complete turn for each 3.6 residues. The helix may be coiled in a right-handed or left-handed direction, but the right-handed helix is the most stable. The α-keratins of hair and wool consist of bundles or cables of 3 or 7 such α-helical coils twisted around each other (Fig. 2–9B).

In other proteins such as **fibroin,** the fibrous protein of silk, the polypeptide chains are in an extended zigzag configuration. These chains are arranged alongside each other to form a **pleated sheet** structure, in which the adjacent polypeptide chains run in opposite directions or are antiparallel to each other (Fig. 2–10). The adjacent chains in the pleated sheet are held together by hydrogen bonds.

FIGURE 2–8   Linus Pauling (1901–   ) Professor of Chemistry, Stanford University. Winner of the 1954 Nobel Prize for Chemistry and the 1962 Nobel Prize for Peace.

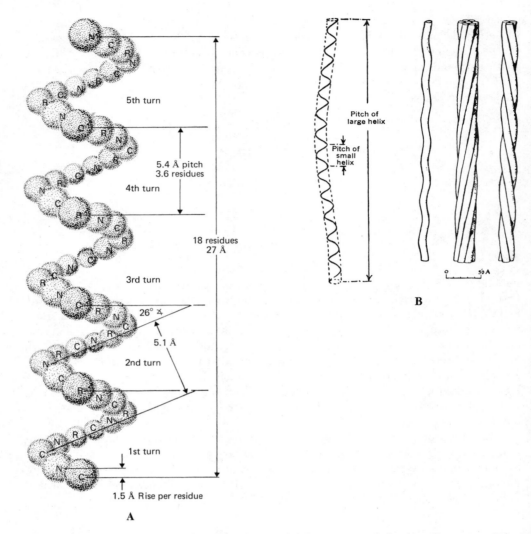

5th turn

5.4 Å pitch
3.6 residues

4th turn

18 residues
27 Å

3rd turn

26° ∢

5.1 Å

2nd turn

1st turn

1.5 Å Rise per residue

A

Pitch of
large helix

Pitch of
small
helix

B

FIGURE 2-9   *A,* Representation of a polypeptide chain as an α-helical configuration. (After Pauling and Corey: Proc. Intern. Wool Textile Research Conf., *B,* 249, 1955.) *B,* Structure of compound α-helices, proposed to explain the structure of α-keratin. **Left,** The coiling of the axis of an α-helix into a long helix. **Right,** Diagrams of a compound α-helix, of a 7-strand α-cable, and a 3-strand α-rope. (From Pauling, in Edsall and Wyman: Biophysical Chemistry, Vol. I. Academic Press, 1958.)

TERTIARY STRUCTURE

The polypeptide chains of globular proteins are more extensively folded or coiled than those of fibrous proteins. This results from the activity of several types of bonds that hold the structure in a more complex and rigid shape. These bonds are responsible for the **tertiary structure** of proteins, and they exert stronger forces than hydrogen bonds in holding together polypeptide chains or folds of individual chains. A strong **covalent bond** is formed between two cysteine residues, resulting in the disulfide bond. **Salt linkages,** or **ionic bonds,** may be formed between the basic amino acid residues of lysine and arginine and the dicarboxylic amino acids such as aspartic and glutamic. Also, there are many examples of **hydrophobic bonding** that result from the close proximity of

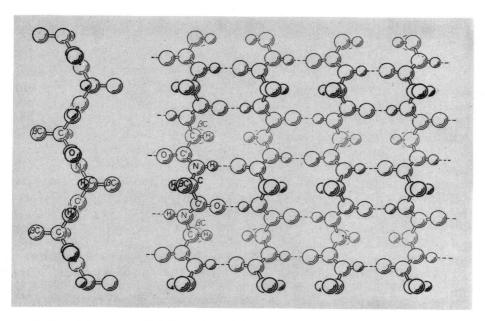

FIGURE 2-10   The antiparallel-chain pleated sheet structure with hydrogen bond arrangement. (From L. Pauling and R. B. Corey: Proc. Nat. Acad. Sci., 37:729, 1951.)

aromatic groups or of like aliphatic groups on amino acid residues. Examples of these bonds may be seen in Figure 2–11.

X-Ray Diffraction Analysis

Sanger's important contribution to protein structure, sequencing the amino acids in the insulin molecule, was matched by the development of the first three-dimensional picture of the protein molecule **myoglobin** from x-ray diffraction data by Kendrew and his co-workers. Myoglobin has a molecular weight of 17,000 and consists of a single polypeptide chain with 153 amino acid residues. It crystallizes readily from muscle extracts of the sperm whale. When x-rays strike an atom, they are diffracted (reflected) in proportion to the number of extranuclear electrons in the atom. The heavier atoms with a high atomic number, therefore, produce more diffraction than lighter atoms. A crystal, such as a protein crystal of myoglobin, when bombarded by a beam of monochromatic x-rays, yields a photographic pattern of the electron density of the atoms in the molecule. From a large series of electron density photographs in different planes, a three-dimensional

FIGURE 2-11   Some types of noncovalent bonds which stabilize protein structure: (*a*) Electrostatic interaction; (*b*) hydrogen bonding between tyrosine residues and carboxyl groups on side chains; (*c*) hydrophobic interaction of nonpolar side chains caused by the mutual repulsion of solvent; (*d*) dipole-dipole interaction; (*e*) disulfide linkage, a covalent bond. (After Anfinsen: The Molecular Basis of Evolution, New York, John Wiley and Sons, 1959, p. 102.)

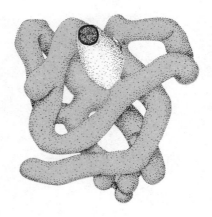

FIGURE 2-12   Model of the myoglobin molecule, derived from the 6 Å Fourier synthesis. The heme group is a dark gray disk (center top). (After Kendrew: Science, *139*:1261, 1963.)

picture of myoglobin was constructed. The model of the myoglobin that resulted from these studies is shown in Figure 2–12. The structural resolution achieved by Kendrew indicated that the major portion of the polypeptide chain was in the form of the right-handed helix proposed by Pauling.

Quaternary Structure

This level of protein structure involves the polymerization, or degree of aggregation, of protein units. The hemoglobin molecule is a good example of subunit structure in proteins. **Native hemoglobin** has a molecular weight of 68,000 in a neutral solution. If the solution is diluted, made acid, or 4M with urea, the molecular weight changes to 34,000. This dissociation is due to the four polypeptide chains which exist in hemoglobin as two pairs of α and β chains. The enzyme **phosphorylase a,** which will be discussed under carbohydrate metabolism, contains four subunits which are inactive as catalysts until they are joined as a tetramer. **Insulin** is another example of a protein hormone containing subunits, and there are several proteins that are split into subunits when their disulfide bonds are converted to sulfhydryl groups. Several enzyme molecules have been separated into isoenzymes by physical techniques such as electrophoresis (see p. 174). **Isoenzymes** apparently have the same activity and molecular weight as the parent enzyme but may be synthesized in different organs in the body. For example, lactic acid dehydrogenase is an enzyme with five isoenzymes, each having a molecular weight of 135,000 and essentially the same enzymatic activity. By treatment with urea each of these molecules was found to be a tetramer with four polypeptide subunits of equal size and two types, A and B. LDH-1 consists of all type B polypeptides, and LDH-5 all type A. When LDH-1 and -5 are mixed in equal proportions, they redistribute into a random mixture of the five isoenzymes. Isoenzymes are becoming increasingly important in diagnosis of disease states.

CLASSIFICATION OF PROTEINS

Proteins are most often classified on the basis of their chemical composition or solubility properties. Of the three main types, **simple proteins** are classified by solubility properties, **conjugated proteins** by the non-protein groups, and **derived proteins** by the method of alteration.

Simple proteins such as **protamines** in the form of salmine and sturine from fish sperm, **histones** in the form of the globin in hemoglobin, and **albumins** in the form of egg albumin and serum albumin are all soluble in water, and protamines and histones contain a high proportion of basic amino acids. The **globulins** as lactoglobulin in milk are insoluble in water but soluble in dilute salt solutions; the **glutelins** as glutenin in wheat are insoluble in water and dilute salt solutions, but are soluble in dilute acid or alkaline solutions; **prolamines** as zein in corn and gliadin in wheat are soluble in 70 to 80 per cent ethyl alcohol; whereas the **albuminoids** as keratin in hair, horn, and feathers are insoluble in all the solvents mentioned above and can be dissolved only by hydrolysis.

Conjugated proteins include **nucleoproteins,** which consist of a basic protein such as histones or protamines combined with nucleic acid. They are found in cell nuclei and mitochondria. **Phosphoproteins** as casein in milk and vitellin in egg yolk are proteins linked to phosphoric acid; **glycoproteins** are composed of a protein and a carbohydrate and occur in mucin in saliva and mucoids in tendon and cartilage, whereas **chromoproteins** such as hemoglobin and cytochromes consist of a protein combined with a colored compound. **Lipoproteins** are proteins combined with lipids such as fatty acids, fats, and lecithin, and are found in serum, brain, and nervous tissue.

Derived proteins are an indefinite type of protein produced, for example, by partial hydrolysis, denaturation, and heat, and are represented by proteoses, peptones, meta-proteins, and coagulated proteins.

DETERMINATION OF PROTEINS

It is often desirable to know the protein content of various foods and biological material. The analysis of the protein content of such material is based on its nitrogen content. Since the average nitrogen content of proteins is 16 per cent, the protein content of a substance may be obtained by multiplying its nitrogen value by the factor $100/16 = 6.25$. For example, if a certain food contained 2 per cent nitrogen on analysis, its protein content would equal 2 times 6.25, or 12.5 per cent.

Since the determination of nitrogen requires considerable equipment and is time-consuming, several colorimetric methods for the quantitative estimation of protein have been developed. Methods based on the biuret test and the ninhydrin test are commonly used when the total protein content of many specimens is required. The aromatic amino acids, especially tryptophan and tyrosine, in proteins absorb ultraviolet light at a wave length of 280 nm. The measurement of light absorption at 280 nm is a convenient method for determining the amount of protein in a solution or in an eluate from a chromatography column.

DENATURATION OF PROTEINS

Denaturation of a protein molecule causes changes in the structure that result in marked alterations of the physical properties of the protein. When in solution, proteins are readily denatured by standing in acids or alkalies, shaking, heating, reducing agents, detergents, organic solvents, and exposure to x-rays and light. Some of the effects of **denaturation** are loss of biological activity, decreased solubility at the isoelectric point, increased susceptibility to hydrolysis by proteolytic enzymes, and increased reactivity of groups that had been masked by the folding of chains in the native protein. Examples of the last-mentioned are the uncovered SH groups of cysteine and the OH groups of tyrosine.

The cleavage of several hydrogen bonds and of several possible disulfide bonds often results in the loss of biological activity by denaturation. In some proteins, the native configuration is so stable that denaturation changes are reversible. Hemoglobin, for example, can be reversibly denatured.

PRECIPITATION OF PROTEINS

One of the most important characteristics of proteins is the ease with which they are precipitated by certain reagents. Many of the normal functions in the body are essentially precipitation reactions: for example, the clotting of blood or the precipitation of casein by rennin during digestion. Since animal tissues are chiefly protein in nature, reagents that precipitate protein will have a marked toxic effect if introduced into the body. Bacteria, which are mainly protein, are effectively destroyed when treated with suitable precipitants. Many of the common poisons and disinfectants act in this way. The following paragraphs contain a brief summary of the most common methods of protein precipitation.

By Heat Coagulation

When most protein solutions are heated, the protein becomes insoluble and precipitates, forming coagulated protein. Many protein foods coagulate when they are cooked. Tissue proteins and bacterial proteins are readily coagulated by heat. Routine examinations of urine specimens for protein are made by heating the urine in a test tube to coagulate any protein that might be present.

By Alcohol

Alcohol coagulates all proteins except the prolamines. A 70 per cent solution of alcohol is commonly used to sterilize the skin, since it effectively penetrates the bacteria. A 95 per cent solution of alcohol is not effective because it merely coagulates the surface of the bacteria and does not destroy them.

By Concentrated Inorganic Acids

Proteins are precipitated from their solutions by concentrated acids such as hydrochloric, sulfuric, and nitric acid. Casein, for example, is precipitated from milk as a curd when acted on by the hydrochloric acid of the gastric juice.

By Salts of Heavy Metals

Salts of heavy metals, such as mercuric chloride and silver nitrate, precipitate proteins. Since proteins behave as zwitterions, they will ionize as negative charges in neutral or alkaline solutions. The reaction with silver ions may be illustrated as follows:

$$R-\underset{\underset{NH_2}{|}}{CH}-COO^- + Ag^+ \rightarrow R-\underset{\underset{NH_2}{|}}{CH}-COOAg$$

Protein   Silver proteinate

These salts are used for their disinfecting action and are toxic when taken internally. A protein solution such as egg white or milk, when given as an antidote in cases of poisoning with heavy metals, combines with the metallic salts. The precipitate that is formed must be removed by the use of an emetic before the protein is digested and the heavy metal is set free to act on the tissue protein. A silver salt such as Argyrol is used in nose and throat infections, and silver nitrate is used to cauterize wounds and to prevent gonorrheal infection in the eyes of newborn babies.

By Alkaloidal Reagents

Tannic, picric, and tungstic acids are common alkaloidal reagents that will precipitate proteins from solution. When in acid solution the protein as a zwitterion ionizes as a positive charge. It will therefore react with picric acid as shown:

$$\text{R}-\overset{\displaystyle |}{\underset{\displaystyle \underset{\text{NH}_3{}^+}{|}}{\text{CH}}}-\text{COOH} + \text{picric acid} \rightarrow \text{R}-\overset{\displaystyle |}{\underset{\displaystyle \underset{\text{NH}_3-\text{picrate}}{|}}{\text{CH}}}-\text{COOH}$$

Protein   Protein picrate

Tannic and picric acids are sometimes used in the treatment of burns. When a solution of either of these acids is sprayed on extensively burned areas, it precipitates the protein to form a protective coating; this excludes air from the burn and prevents the loss of water. In an emergency, strong tea may be used as a source of tannic acid for the treatment of severe burns. Many other therapeutic agents have been used in the treatment of burns, the most recent being penicillin. Nevertheless, considerable quantities of tannic and picric acid preparations are still employed for this purpose.

By Salting Out

Most proteins are insoluble in a saturated solution of a salt such as ammonium sulfate. When it is desirable to isolate a protein from a solution without appreciably altering its chemical nature or properties, the protein may be precipitated by saturating the solution with $(NH_4)_2SO_4$. After filtration, the excess $(NH_4)_2SO_4$ is usually removed by dialysis. This salting out process finds wide application in the isolation of biologically active proteins.

IMPORTANT TERMS AND CONCEPTS

acidic amino acids
alpha amino acids
alpha helix
amino acid sequence
aromatic amino acids
basic amino acids
chromatography
heterocyclic amino acids
hydrogen bonds

isoelectric point
isoenzymes
peptide linkage
pK of amino acids
polypeptides
primary structure
secondary structure
tertiary structure
zwitterions

QUESTIONS

1. List the five main elements present in proteins with their average content. Using this list, how would you differentiate proteins from carbohydrates and fats?

2. What products are formed by complete hydrolysis of proteins? Which chemical agents would be used for complete hydrolysis?

3. Illustrate the ionized and undissociated form of an amino acid. Why are the ionized forms important?

4. Write the formula for an alpha amino acid containing three carbon atoms. How would you name this amino acid?

5. Write the formula and name of an amino acid that contains a heterocyclic ring.

6. Write the formula of an amino acid as a zwitterion and use the structure to explain the isoelectric point.

7. Illustrate with equations two reactions that involve the amino group of an amino acid.

8. Given the dipeptide alanylglycine as an unknown, explain the procedures you would employ to:
   a. Prove the identity of the two amino acids.
   b. Determine which amino acid in the dipeptide contained the free amino group.

9. Why is the biuret reagent used in the quantitative analysis of protein and the ninhydrin reagent used to spray amino acid spots in chromatography? Explain.

10. Explain how Sanger split the two chains of the insulin molecule apart and determined their N-terminal amino acids.

11. Outline the amino acid sequence of vasopressin. How could the position of the disulfide bridge in the molecule be established?

12. What are the major characteristics of the secondary structure of proteins?

13. How does the primary structure differ from the tertiary structure of a protein? Explain.

14. What is the relationship between isoenzymes and protein structure? Explain.

15. What is the basis of the classification of the different types of proteins?

16. Describe two methods that can be used to determine the amount of protein in a solution.

17. Why is a protein solution used as an antidote in cases of poisoning with heavy metals? Explain.

18. Illustrate with equations the precipitation of proteins with tannic acid and with silver salts.

Chapter 3

NUCLEIC ACIDS

The *objectives* of this chapter are to enable the student to:

1. Illustrate the structure of the products of complete hydrolysis of a nucleoprotein.
2. Distinguish between the components in DNA and RNA.
3. Recognize the difference between ATP and dATP.
4. Illustrate the tetranucleotide portion of one chain of DNA.
5. Illustrate hydrogen bonding and antiparallel chains in the DNA molecule.
6. Distinguish between the different types of RNA molecules.

Nucleic acids were first isolated from the nuclei of cells of salmon sperm by Miescher in 1870. The acidic substance he extracted was called **nuclein.** At first nucleic acids were thought to be fairly simple compounds that were conjugated with proteins to form nucleoproteins. These proteins are characterized by their relatively high content of basic amino acids such as arginine and lysine. Protamines and histones are examples of such proteins. We know now that nucleic acids are polymers of large molecular weight with nucleotides as the repeating unit. Deoxyribonucleic acid (DNA) is present in the nucleus and ribonucleic acid (RNA) in the cytoplasm of all living cells. DNA and RNA are essentially responsible for the transmission of genetic information and the synthesis of protein by the cell, respectively. Progressive hydrolysis of a nucleoprotein would yield the protein and nucleic acid and its components as shown:

$$\text{Nucleoprotein} \rightarrow \text{Nucleic acid} + \text{Protein}$$
$$\downarrow$$
$$\text{Nucleotides}$$
$$\downarrow$$
$$\text{Nucleosides} + H_3PO_4$$
$$\downarrow$$
$$\text{Purines and Pyrimidines} + \text{Pentose}$$

More specific information concerning the hydrolysis products of DNA and RNA is necessary before the structure of these nucleic acids is considered.

DNA	RNA
Adenine	Adenine
Guanine	Guanine
Cytosine	Cytosine
Thymine	Uracil
D-2-Deoxyribose	D-Ribose
H_3PO_4	H_3PO_4

THE PYRIMIDINE AND PURINE BASES

The heterocyclic rings that form the nucleus for both the pyrimidine and purine bases are shown below:

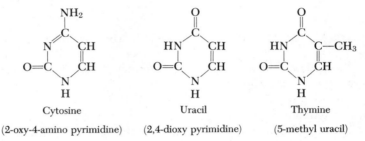

The pyrimidine bases found in nucleic acids include cytosine, uracil, and thymine, which are represented as follows:

Cytosine	Uracil	Thymine
(2-oxy-4-amino pyrimidine)	(2,4-dioxy pyrimidine)	(5-methyl uracil)

Both DNA and RNA contain the purines adenine and guanine.

Adenine
(6-amino purine)

Guanine
(2-amino-6-oxy purine)

THE PENTOSE SUGARS

The pentose sugars that are essential components of the nucleic acids are discussed on pages 59 and 61. On hydrolysis RNA yields β-D-ribose, whereas DNA contains β-2-deoxy-D-ribose. Both pentose sugars occur in the furanose form.

β-D-Ribose

β-2-deoxy-D-Ribose

NUCLEOSIDES

When a purine or pyrimidine base is combined with β-D-ribose or β-2-deoxy-D-ribose, the resultant molecule is called a **nucleoside.** The linkage of the two components is from the nitrogen of the base (position 1 in pyrimidines, 9 in purines) to carbon 1 of the pentose sugars. The formation of a nucleoside involves a reaction between the pyrimidine or purine and ribose phosphate. A nucleophilic displacement reaction occurs, for example, in which the entering pyrimidine group takes its position on the opposite face of the ribose molecule from which the phosphate has left. Important examples of nucleosides are cytidine and adenosine:

Cytidine   Adenosine

NUCLEOTIDES

When a phosphoric acid is attached to a hydroxyl group of the pentose sugar in the nucleoside by an ester linkage, the result is a **nucleotide.** The esterification may occur on the 2′, 3′, or 5′ hydroxyl of ribose and the 3′ or 5′ hydroxyl of deoxyribose. In the nucleotides, the carbon atoms of ribose or deoxyribose are designated by prime numbers to distinguish them from the atoms in the purine or pyrimidine bases. Yeast nucleic acid contains four mononucleotides: adenylic acid, guanylic acid, cytidylic acid, and uridylic acid. All these compounds include D-ribose in their structure and are named as acids because of the ionizable hydrogens of the phosphate group. Thymus nucleic acid yields nucleotides on hydrolysis that contain β-2-deoxy-D-ribose and thymidylic acid instead of uridylic acid. These structures may be represented by adenylic acid and uridylic acid, two compounds of prime importance in muscle metabolism and carbohydrate metabolism:

Adenylic acid
(adenosine-5′-monophosphate, AMP)

Uridylic acid
(uridine-3′-monophosphate, UMP)

In addition to the nucleotides that are integral components of yeast and thymus nucleic acids, several nucleotides and their derivatives occur free in the tissues and are essential constituents of tissue metabolism. For example, adenylic acid, also designated adenosine monophosphate (AMP), is found in muscle tissue. The biochemist commonly uses abbreviations such as AMP and, for uridine monophosphate, UMP, to designate nucleotides. Adenylic acid may also exist as the diphosphate (adenosine diphosphate, ADP) or as the triphosphate (adenosine triphosphate, ATP) (p. 105). These two compounds are sources of high energy phosphate bonds and are involved in many metabolic reactions. The nucleosides and nucleotides also often contain β-2-deoxy-D-ribose as their constituent sugar. The mono-, di-, and triphosphate compounds illustrated below would be designated dAMP, dADP, and dATP if they contained deoxyribose.

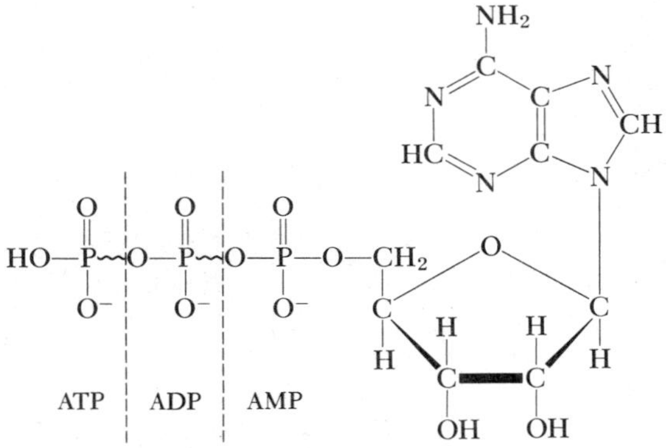

Adenosine triphosphate

Nucleotides also combine with vitamins to form coenzymes, which will be discussed in Chapter 7.

TOPIC OF CURRENT INTEREST

THE DOUBLE HELIX OF DNA

As is often the case in biochemical developments, thousands of hours of research were required to establish the background information essential to the proposal by Watson and Crick (Fig. 3–1) of the double helical structure of DNA. One of the earliest observations relating to the structure of DNA was the isolation of adenosine-3'-monophosphate from yeast RNA by Levene in 1918. Later it was shown that mild acid hydrolysis of DNA would yield 3',5'-diphosphate derivatives of thymidine. This, along with other experimental evidence, suggested that nucleic acids were composed of nucleotides joined together by a diester linkage between the 3' position of one ribose molecule and the 5' position of the adjacent nucleotide (Fig. 3–2). A major obstacle in the early studies of DNA structure involved the preparation of DNA molecules representative of those occurring naturally in the cell. Treatment with acid or alkali separated the DNA from other cellular components, but markedly changed the native structure. Separation schemes using neutral salt solutions were eventually successful in producing samples of thymus gland DNA with molecular weights of several million that were very similar to DNA in the cellular state. With these and similar preparations, Chargaff and his co-workers, using ion exchange chromatography, observed in 1950 that the sum of the purine nucleotides equaled the sum of the pyrimidine nucleotides, and that the adenine content was equal to the thymine content and the guanine content equal to that of cytosine. The latter observation was of basic importance, since it suggested that adenine and thymine nucleotides could be paired structurally and joined by two hydrogen bonds, whereas cytosine and guanine could form structural pairs joined by three hydrogen bonds. Titration data suggested

FIGURE 3–1   a, James Watson (1928–   ) Professor of Biology, Harvard University. Shared the Nobel Prize for Medicine and Physiology in 1962 with Crick and Wilkins. Proposed double helical structure of DNA and role of RNA in protein synthesis.   b, Francis Crick (1916–   ) member British Medical Research Council, Laboratory on Molecular Biology. Shared the Nobel Prize for Medicine and Physiology in 1962 with Watson and Wilkins. Proposed model for double helical structure of DNA.

FIGURE 3–2   The tetranucleotide portion of one chain of DNA, above, and a shorthand representation of the structure, below.

35

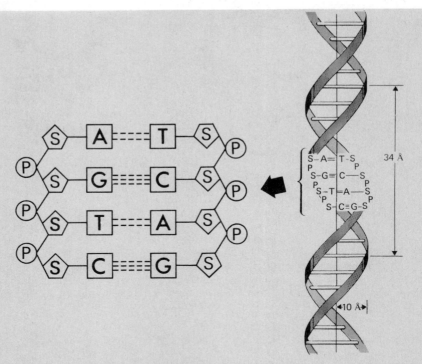

FIGURE 3-3   Double helix of DNA. Here P means phosphate diester, S means deoxyribose, A=T is the adenine-thymine pairing, and G≡C is the guanine-cytosine pairing. (After Conn and Stumpf: Outlines of Biochemistry, 2nd Ed. New York, Wiley, 1963, p. 108.)

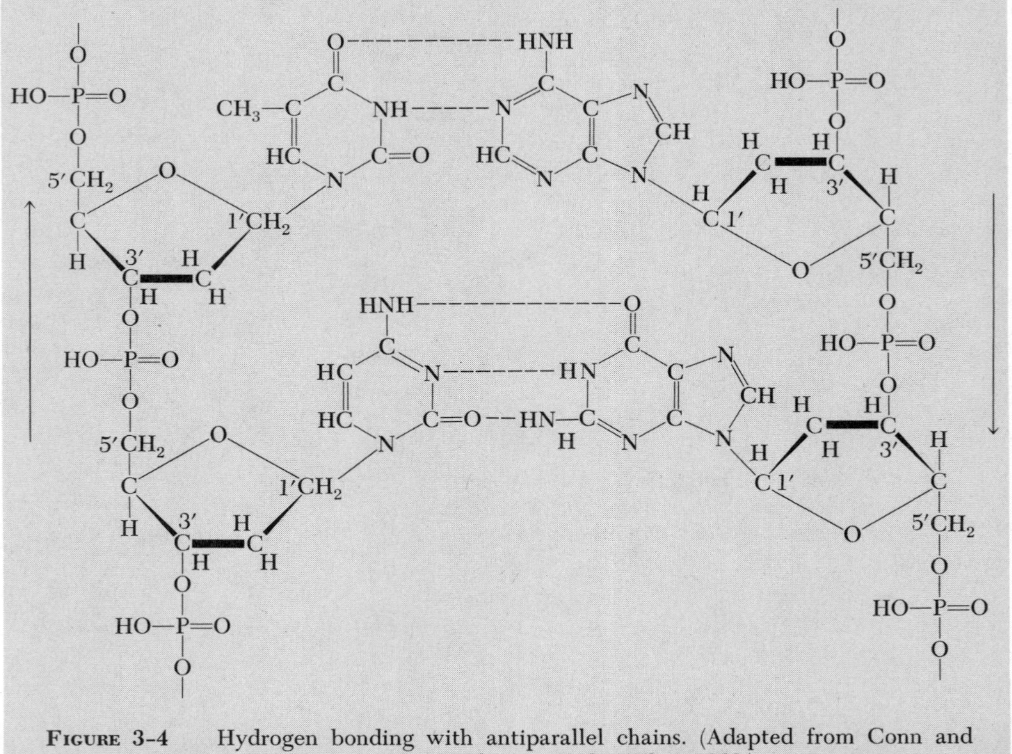

FIGURE 3-4   Hydrogen bonding with antiparallel chains. (Adapted from Conn and Stumpf: Outlines of Biochemistry, 2nd Ed. New York, Wiley, 1963.)

the presence of long nucleotide chains held together by hydrogen bonding between base residues.

The experimental evidence gained from all of these studies set the stage for the final breakthrough, in which Wilkins observed that DNA from different sources had very similar x-ray diffraction patterns. A uniform molecular pattern for all DNA was therefore probable. Wilkins' studies in 1952 also suggested that DNA consisted of two or more polynucleotide chains arranged in a helical structure. Once all this data was available, it became a matter of time before a group of investigators would first propose a workable theory of DNA structure satisfactory to all scientists. Watson and Crick constructed their model in 1953 to account for all the known properties of DNA. In the model, two polynucleotide chains were twisted into a helix (Fig. 3–3). The chains were composed of deoxyribose phosphates joined together by 3′,5′-diester linkages, with the bases projecting perpendicularly from the chain into the center axis. To fit the base-pairing evidence, it was necessary that for each adenine projecting from the central axis, one thymine must project toward the adenine from the second chain of the helix and be held by two hydrogen bonds to the adenine. Cytosine and guanine exhibited similar behavior in a different area of the helix, and were joined by three hydrogen bonds. This resulted in a spatial structure of two polynucleotide chains coiled around a common axis and held together by the hydrogen bonding of adenine with thymine and cytosine with guanine. To achieve this spatial arrangement and base-pairing, it was necessary for the chains to be exact complements of each other; *i.e.*, instead of running in the same direction, they were antiparallel chains of polynucleotides (Fig. 3–4).

A most dramatic confirmation of the Watson and Crick model for DNA was made by Kornberg, who isolated from *E. coli* an enzyme that catalyzed the synthesis of DNA from the four deoxyribose nucleotide triphosphates in the presence of Mg^{++} and a small amount of DNA, which served as a primer for the synthesis. He found that the synthesized DNA had the same base composition as the primer DNA, regardless of the relative quantities of dATP, dGTP, dCTP, and dTTP used in the reaction mixture.

For their successful proposal of the structure of DNA, Watson and Crick were awarded the Nobel Prize in 1962.

NUCLEIC ACIDS

The Structure of RNA

The RNA molecules in the cell are of three major types: ribosomal RNA (r-RNA), messenger RNA (m-RNA), and transfer RNA (t-RNA). All three types have single strands of polynucleotides (in contrast to double-stranded DNA), and each type occurs in multiple molecular forms. Ribosomal RNA occurs in at least three forms, transfer RNA in about sixty forms, and messenger RNA in several hundred molecular forms. In *E. coli* cells, most of the RNA is in the cytoplasm, but in eukaryotic cells such as rat liver, the RNA is distributed in the nucleus, ribosomes, mitochondria, and cytoplasm. The largest molecules are **ribosomal RNA** (r-RNA), with molecular weights of a few million. They are associated with the structure of the ribosomes and serve as a template for protein synthesis within the cytoplasm. Ribosomal RNA is the predominant RNA in the cell, since it comprises 75 to 80 per cent of the total RNA in liver cells and *E. coli*. **Messenger RNA** (m-RNA) molecules are varied in size, with molecular weights from 300,000 to two million. The m-RNA molecules carry the genetic message from the DNA in the nucleus to the protein-synthesizing sites. The smallest RNA molecules are called **transfer RNA** (t-RNA), and have molecular weights from 25,000 to 40,000. Their function is to transport specific amino acids to their specific sites on the protein-synthesizing template. The basic composition of three transfer RNAs is known. They are composed of single strands of nucleotides bent into

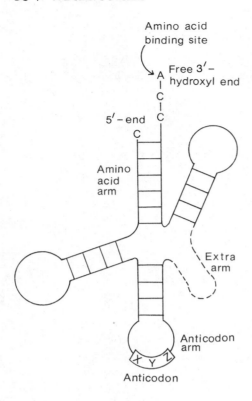

Amino acid
binding site

Free 3'–
hydroxyl end

A
C
C
5'–end C
C

Amino
acid
arm

Extra
arm

Anticodon
arm

X Y Z

Anticodon

FIGURE 3–5   Cloverleaf structure of t-RNAs. This configuration assumes maximal base pairing. Some t-RNAs have an extra arm as shown. (Adapted from Lehninger: Short Course in Biochemistry, New York, Worth Publishing, Inc., 1973, p. 406.)

cloverleaf-type structures to give the maximum number of hydrogen-bonded pairs (Fig. 3–5). The closed loop of the chain at one end of the cloverleaf contains a sequence of three bases that serves as an anticodon for a specific amino acid (see Chapter 12). The role of these RNA molecules will be discussed in Chapter 12.

THE BIOLOGIC IMPORTANCE OF THE NUCLEIC ACIDS

Originally RNA was associated only with yeast and was thought to be restricted to plant sources. DNA from thymus tissue represented the nucleic acids of animal tissues. As methods for their determination have been developed, both RNA and DNA have been found in practically all types of cells. DNA appears to be restricted to the nucleus, most specifically to the chromosomes, whereas RNA occurs both in the cytoplasm and nucleus of a cell. The overall function of DNA and its relation to RNA and protein synthesis are outlined in Figure 3–6. A more detailed explanation of the role of DNA and RNA in protein synthesis will be discussed in Chapter 12. The biological importance of the nucleic acids in genetics is stressed in Chapter 13.

Synthesis of Nucleic Acids

From the composition of nucleic acids as shown on pages 35–36, it is apparent that the synthesis would involve the copolymerization of four ribonucleotides for RNA and four deoxyribonucleotides for DNA. The DNA molecule has been synthesized by Kornberg and his co-workers by treating a mixture of deoxyribonucleotides with an enzyme isolated from bacteria. In the presence of all four deoxyribonucleotides, a DNA primer, and Mg^{+2}, the enzyme **DNA polymerase** can synthesize DNA. Enzymes called **RNA polymerases** or

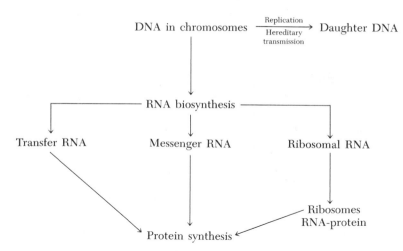

FIGURE 3-6   The relation of DNA to RNA in protein synthesis.

transcriptases found in plants, animals, and bacteria catalyze the synthesis of RNA from ribonucleoside triphosphates, ATP, GTP, CTP, and UTP. The synthesis is DNA-dependent, and requires Mg^{+2}, in addition to all four ribonucleoside triphosphates. The DNA serves as a template, and the base sequence in the DNA is transcribed into a corresponding sequence in the RNA. These nucleic acid molecules are similar to those from natural sources and are used to study the properties, compositions, and reactions of these large molecules.

Multiplication of viruses within cells can occur by a similar process in which the viral genes consist of DNA that transmits information to the RNA of the cell and then into the cell proteins. Several common viruses which cause poliomyelitis, the common cold, and influenza are called RNA viruses, since the RNA replicates directly into new copies of RNA and translates information directly to proteins of the cell without DNA involvement in their replication. Recently evidence for a reverse flow of genetic information from RNA to DNA has been obtained in experiments using the Rous sarcoma virus. It has been suggested that in normal cells there are regions of DNA that serve as templates for the synthesis of RNA, and that this RNA serves in turn as a template for the synthesis of DNA which then becomes integrated with the cellular DNA. These experiments may be valuable in an explanation of the origin of cancer in man. Since it was shown that cancer-causing RNA viruses can produce a DNA transcript of the viral RNA, it is possible that the viral RNA may transmit genetic information to the genes of cells that will eventually surface as spontaneous cancer.

IMPORTANT TERMS AND CONCEPTS

adenine
antiparallel chains
cytosine
DNA
double helix
guanine
hydrogen bonding
nucleoside

nucleotide
purine
pyrimidine
replication
RNA
tetranucleotide
thymine
uracil

QUESTIONS

1. What type of protein is found in nucleoproteins? What products result from complete hydrolysis of nucleic acids?

2. What are the two major types of nucleic acids? List the composition of each type.

3. Write the formula and name for a nucleoside containing β-2-deoxyribose.

4. What is the difference between ATP and ADP; ADP and AMP? Illustrate these differences with a composite formula.

5. How are the nucleotides linked together chemically when they form nucleic acids?

6. Prepare a sketch of the DNA molecule showing the double helix and the hydrogen bonding between the bases.

7. What is meant by the antiparallel chain structure of DNA?

8. Illustrate a tetranucleotide portion of a DNA molecule in a simplified fashion and explain the notations used.

9. Briefly describe the three major types of RNA.

10. Briefly discuss the synthesis of RNA.

Chapter 4

ENZYMES

The *objectives* of this chapter are to enable the student to:

1. Give a definition for an enzyme and describe its chemical nature.
2. Recognize the systematic code number of a common enzyme.
3. Explain the value of the Michaelis constant, K_m, of an enzyme.
4. Recognize the relationship between the active site of an enzyme and its specificity of action.
5. Describe the effect of pH, temperature, and end products on an enzyme reaction.
6. Recognize and illustrate the difference between a competitive and a noncompetitive inhibitor.

Living cells function as effective chemical machines because they contain enzymes, which are catalysts capable of greatly increasing the rate of specific chemical reactions. Nearly 2000 different enzymes are known, and it is estimated that there are as many as 1000 separate enzymes in a single cell. They differ from inorganic catalysts in their reaction specificity, efficiency, and their capacity to operate under the mild temperature conditions and hydrogen ion concentration found in the living cell. Enzymes combine with substrate molecules during the catalytic process in such a way that the active site of the enzyme molecule fits the substrate with a nearly perfect lock and key correspondence. Also, during the process the complex of bound enzyme and substrate assumes a new configuration in which the bound substrate is modified so that it becomes a new compound upon release from the enzyme; the enzyme then resumes its original form. The hundreds of chemical reactions in a cell that are catalyzed by enzymes do not occur independently of each other, but are linked into sequences of consecutive reactions that have common intermediates. Often the product of one reaction becomes the substrate or reactant of another reaction.

THE CHEMICAL NATURE OF ENZYMES

Enzymes have always been considered as catalysts and are often compared to inorganic catalytic agents such as platinum and nickel. These inorganic agents are often used in conjunction with high temperatures, high pressures, and favorable chemical conditions. Few of these conditions occur when an enzyme reacts in body tissue, at body

*Gibbs'
Free
Energy*

temperature, and at the pH of body fluids. Enzymes were originally defined as catalysts, organic in nature, formed by living cells, but independent of the presence of the cells in their action. A more current definition would state that *enzymes are proteins, formed by a living cell, which catalyze a thermodynamically possible reaction by lowering the activation energy so the rate of reaction is compatible with the conditions in the cell.* An enzyme does not change the $\Delta G°$ or equilibrium constant of a reaction.

Enzymes are proteins composed of the same 22 amino acids found in other proteins. Their characteristic shapes are dictated by the sequence of amino acids. Because they are proteins, the reactive groups of enzymes are spaced in rigidly defined positions, and the geometric specificity of their active sites produces a degree of control over the catalytic process that is not characteristic of inorganic catalysts. The purification, crystallization, and inactivation procedures exactly parallel those for pure proteins. For example, excessive heat, alcohol, salts of heavy metals, and concentrated inorganic acids will cause coagulation or precipitation of the protein material and thus inactivate an enzyme.

The naming of enzymes has become increasingly complex as many new specific enzymes have been described. Originally they were named according to their source or according to the method of separation when they were discovered. As the family of enzymes grew, they were named in a more orderly fashion by adding the ending -**ase** to the root of the name of the substrate. An enzyme's **substrate** is the compound or type of substance upon which it acts. For example, sucrase catalyzes the hydrolysis of sucrose, lipase is an enzyme that hydrolyzes lipids, and urease is the enzyme that splits urea. This system also was used to name types of enzymes such as proteases, oxidases, and hydrolases. The discovery of so many enzyme mechanisms in the past few years has resulted in a mass of complex substrates and enzyme nomenclature. The problem was assigned to a Commission on Enzymes of the International Union of Biochemistry, whose members studied the system of nomenclature for six years before publishing a report in 1961. They were not in favor of eliminating all the names in common usage, but recommended the use of two names for an enzyme. One was the trivial name, either the one in common use or a simple name describing the activity of the enzyme. The other was constructed by the addition of the ending -**ase** to an accurate chemical name for the substrate.

They also devised a systematic code number (E.C.) for each enzyme. This number characterizes the type of reaction catalyzed by the enzyme as to class (first number), subclass (second number), and sub-subclass (third number). The fourth number is specific for the particular enzyme named. As an example, pancreatic lipase is assigned the number 3.1.1.3, which specifies a hydrolase (3) that acts on ester bonds (3.1) that are carboxylic esters (3.1.1.), with the specific number for the lipase (3.1.1.3).

CLASSIFICATION OF ENZYMES

The current classification of enzymes is based on both the type of chemical reaction catalyzed and the numbering system devised by the Commission. The six major classes of enzymes, with examples of their subclasses and their number codes, are shown in the following classification.

1. *Oxidoreductases.* Enzymes in this group are involved in physiological oxidation processes. An example would be alcohol:NAD oxidoreductase or alcohol dehydrogenase, designated 1.1.1.1, acting on a CHOH group (subclass), with NAD as the coenzyme (sub-subclass).
2. *Transferases.* These enzymes catalyze the transfer of a chemical group from one substrate to another. The transfer of amino, methyl, alkyl, and acyl groups and groups containing phosphorus or sulfur is catalyzed by these enzymes. Enzyme

number 2.1.1.1 is a one-carbon transferase (2.1) that transfers methyl groups (2.1.1), and is specifically named adenosylmethionine:nicotinamide methyltransferase (2.1.1.1).

3. *Hydrolases.* This large group of enzymes catalyzes hydrolytic reactions and includes digestive enzymes such as amylase, sucrase, lipase, and the proteases. A specific example may be lipase 3.1.1.3, described in the preceding section.

4. *Lyases.* These enzymes catalyze the removal of chemical groups without hydrolysis. These enzymes act on C—C, C—O, C—N, and C—S bonds, and include 2-oxoacid carboxy-lyase or pyruvate decarboxylase (4.1.1.1), which acts on a C—C bond (4.1) removing a carboxyl group (4.1.1).

5. *Isomerases.* This group of enzymes catalyzes isomerization reactions. Examples are *cis-trans* isomerases, racemases, and epimerases. An example related to the process of vision (see p. 99) is retinal isomerase (5.2.1.3), in which 5.2 designates a *cis-trans* isomerase and 5.2.1 denotes a *cis-trans* isomerase acting on polyunsaturated hydrocarbons.

6. *Ligases.* These enzymes catalyze the linking together of two molecules with the breaking of a pyrophosphate bond of ATP or a similar triphosphate. An example of a ligase involved in the synthesis of protein would be tyrosine:t-RNA ligase or tyrosyl-t-RNA synthetase, 6.1.1.1, which links two molecules together forming C—O bonds (6.1) or, more specifically, an amino acid-RNA ligase forming C—O bonds (6.1.1).

PURIFICATION OF ENZYMES

Crude extracts of enzymes from tissues or cells may be obtained by grinding the tissue in metal grinders or between ground glass surfaces, by alternate freezing and thawing, or by exposure to ultra-sound. The enzymes may be separated from the extracts by protein precipitation, adsorption on ion exchange resins, electrophoresis (adsorption assisted by the passage of an electrical current), and extraction with various solvents. The first enzyme to be obtained in crystalline form was urease, which was crystallized by Sumner in 1926. Since then about 75 other enzymes have been crystallized.

As enzyme preparations are purified and crystallized they naturally increase their potency of action. The activity of an enzyme, or more specifically one unit of an enzyme, is defined as that amount which will catalyze the transformation of 1 micromole of substrate per minute. A **micromole** is 1 millionth of a gram molecular weight. The potency, or **specific activity,** is expressed as units of enzyme per milligram of protein; the **molecular activity** is defined as units per micromole of enzyme at optimal substrate concentration. To compare the relative activity of different enzymes the **turnover number** is used. This is defined as the number of moles of substrate transformed per mole of enzyme per minute at a definite temperature. Turnover numbers vary from about 10,000 to 5,000,000, with the enzyme catalase exhibiting the highest activity.

PROPERTIES OF ENZYMES

Specificity of Action

Perhaps the major difference between the classical inorganic catalysts, such as platinum and nickel, and enzymes is the specificity of action of the latter. Platinum, for example, will act as a catalyst for several reactions. Enzymes may exhibit different types of specificities, as listed on p. 44.

1. Absolute specificity
2. Stereochemical specificity
3. Reaction, or linkage, specificity
4. Group specificity

Urease exhibits **absolute specificity** in action in that it catalyzes the splitting of a single compound, urea. Other enzymes exhibit **stereochemical specificity;** D-amino acid oxidase is specific for D-amino acids and will not affect the natural L-amino acids. Arginase catalyzes the hydrolysis of L-arginine to urea and ornithine, but will not act on the D-isomer. The comparison between the specificity of enzymes and inorganic catalysts may need reconsideration as more information concerning organometallic catalysts becomes available. For example, optically active rhodium catalysts have been developed that catalyze the synthesis of optically active products.

In Chapter 2 enzymes that split particular bonds between amino acids in peptide fragments were used to help determine the amino acid sequence in proteins. These enzymes are **linkage-specific;** for example, trypsin splits peptide bonds adjacent to lysine or arginine residues, whereas chymotrypsin splits bonds next to aromatic amino acids, such as tyrosine in a polypeptide chain. Enzymes that exhibit **group specificity** are also valuable in sequence studies. Carboxypeptidase is specific for terminal amino acids containing a free carboxyl group, and aminopeptidase splits terminal amino acids with a free amino group off the end of a peptide chain. In general, the specificity of action accounts for the large numbers of enzymes found in cells and tissues, and for the fact that enzymes are involved in all the metabolic reactions that occur in the cell.

Energy of Activation

Cellular processes in biochemistry are designed to obtain energy from food and to transform it to forms of energy useful in body growth and maintenance. It is important to remember that chemical reactions may either release energy, or require energy to proceed. The first type of reaction is characterized by a negative ΔG, with energy content of the reactants greater than that of the products, and a tendency toward spontaneous reaction. The latter type of reaction has a positive ΔG, with less energy content in the reactants and no tendency toward spontaneous reaction. Since all chemical reactions tend to proceed in the direction of forming products of a lower energy content than the reactants, the problem is how to initiate the reaction and make it proceed spontaneously. In the reaction

$$A + B \rightarrow \text{Activated state} \rightarrow C + D$$

energy is required to produce the activated state before the reaction can proceed to products C + D. The energy of activation, E_a, required to change A + B to the activated state is affected by a catalyst. Figure 4–1 illustrates the effect of an enzyme on the energy of activation compared to the effect of the common catalyst, platinum, on the decomposition of hydrogen peroxide to water and oxygen.

To decompose hydrogen peroxide to water and oxygen without a catalyst requires 18.0 kcal/mole. An effective metallic catalyst such as platinum lowers the activation energy of the reaction to 12.0 kcal/mole, compared to the low activation energy required by the enzyme catalase of about 3.0 kcal/mole. In general, enzymes lower activation energies of reactions to the point where they can be readily carried out at body temperature under the conditions of living tissue.

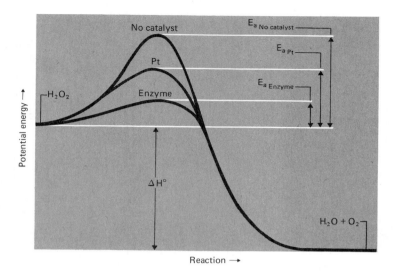

Figure 4-1 Effect of catalysts on the energy of activation in the decomposition of hydrogen peroxide.

ENZYME ACTIVITY

As discussed above, enzymes lower the activation energy and speed up reactions by forming a complex with the substrate. This shifts to a new conformation in which the bonds of the substrate are strained into more reactive positions. This new, more active intermediate complex can shift back to the original enzyme-substrate complex or be converted to products of the reaction plus the original enzyme molecule, as shown here:

$$E \; + \; S \; \rightleftharpoons \; E\text{-}S \; \rightleftharpoons \; ER \; \rightleftharpoons \; E \; + \; P$$

| Enzyme | Substrate | Enzyme-substrate complex | Active intermediate complex | Enzyme | Products |

The activity may be measured by following the chemical change that is catalyzed by the enzyme. The substrate is incubated with the enzyme under favorable conditions, and samples are withdrawn at short intervals for analysis of the end products or analysis of the decrease of substrate concentration. The enzyme lipase, for example, catalyzes the hydrolysis of fat molecules to fatty acids and glycerol. A simple method of measuring the activity of lipase would involve a determination of the rate of appearance of fatty acid molecules.

Effect of Substrate

The concept of the **enzyme-substrate complex** as a transition state in enzyme reactions was first expressed by Michaelis and Menten in 1913. The simple equation they proposed would apply to an enzyme containing a single active site that combines with only one substrate molecule.

$$E \; + \; S \; \underset{k_{-1}}{\overset{k_1}{\rightleftharpoons}} \; ES \; \overset{k_2}{\longrightarrow} \; E \; + \; P$$

| Enzyme | Substrate | Enzyme-substrate complex | Products |

When conditions are arranged so the concentration of the product is kept very low, its tendency to occupy the active site in place of the substrate is negligible. Under these conditions the velocity of the reaction varies with the concentration of the substrate, as

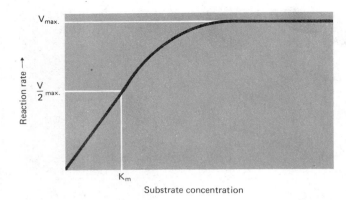

$V_{max.}$

$\frac{V}{2}\,_{max.}$

Reaction rate →

K_m

Substrate concentration

FIGURE 4-2 Effect of substrate concentration on the reaction rate when the enzyme concentration is held constant.

illustrated in Figure 4–2. The velocity of the reaction approaches a plateau value as the substrate concentration is increased. The maximum velocity ($v_{max.}$) is reached when the enzyme is saturated with substrate, a situation that occurs when a molecule of substrate combines immediately with the active site of the enzyme molecule at the moment it releases the product of the reaction. Michaelis and Menten expressed the relation between the actual velocity of an enzyme reaction and the maximum velocity as follows:

$$v = \frac{v_{max.}\,[S]}{K_m + [S]}$$

where K_m is a constant known as the Michaelis constant and [S] is the concentration of the substrate. They demonstrated that the K_m is equal to the substrate concentration at which the velocity of the reaction is half of its maximum (Figure 4–2). This can be shown from the equation, where if $[S] = K_m$, then

$$v = \frac{v_{max.}\,K_m}{K_m + K_m} = \frac{v_{max.}\,K_m}{2\,K_m} = \frac{v_{max.}}{2}$$

A high value of K_m would indicate a low affinity of an enzyme for its substrate, whereas a low K_m (0.001 M, for example) would indicate a high affinity, the active site of the enzyme being half-saturated when its substrate is present at that concentration.

In the cell there is often a balance between the actual substrate concentration and the K_m of the enzyme. Many enzymes appear to be synthesized with K_m values approaching the cellular concentration of their substrates, assuring their effectiveness as catalysts and the self-regulation of the reaction. In situations in which the K_m of an enzyme is very high, the enzyme will exhibit maximum sensitivity to changes in cell substrate concentration, but will be operating at low efficiency. Conversely, if the K_m of an enzyme is very low it will be almost constantly saturated with its substrate, and any variations in substrate concentration will have little effect on the rate of reaction.

The curve shown in Figure 4–2 represents the ideal relationship between the substrate concentration [S] and the maximum velocity $v_{max.}$. Experimentally it is difficult to reproduce this curve for a particular enzyme to obtain an accurate measure of $v_{max.}$ or K_m. By inverting the Michaelis-Menten equation and expressing it as an equation for a straight line, the following relationship may be obtained:

$$\frac{1}{v} = \frac{K_m}{v_{max.}} \times \frac{1}{[S]} \times \frac{1}{v_{max.}}$$

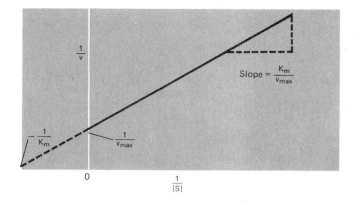

FIGURE 4-3   The Lineweaver-Burk plot of an enzyme reaction.

This equation may then be used to construct a double-reciprocal plot of $\frac{1}{v}$ versus $\frac{1}{[S]}$ to obtain a graphic evaluation of K_m and v_{max}. This is called a Lineweaver-Burk plot, after the investigators who proposed its use (Fig. 4–3). Experimentally, it requires only a few points on the curve to determine K_m; therefore, the Lineweaver-Burk plot method is most often used for this purpose in the laboratory. In addition to the K_m representing a measure of the affinity of an enzyme for its substrate, it also is of practical value in the assay of enzymes. At a substrate concentration of 100 times the K_m value, the enzyme will exhibit a maximum rate of activity or v_{max}. The K_m value, therefore, determines the amount of substrate to use in an enzyme assay.

TOPIC OF CURRENT INTEREST

THE ACTIVE SITE OF ENZYMES

In view of the specificity of action of enzymes, it is reasonable to assume that only a small portion of the enzyme protein is involved in its catalytic activity. The portion of the enzyme molecule to which the substrate binds is called the **active site.** Studies of substrate specificity indicate that the substrate must have a susceptible chemical bond that can be attacked by the enzyme. In addition, some other structural feature is required for its binding to the active site of the enzyme; this feature is probably essential to position the substrate molecule in the proper geometric relationship so that the susceptible bond can be attacked. As an example, acetylcholine is the substrate for the enzyme acetylcholinesterase. The susceptible bond is the ester linkage between choline and the acetyl group; the part of the molecule required for its positioning on the active site is the positively charged quaternary ammonium group adjoining a nonpolar group in the molecule.

Digestive enzymes such as pepsinogen, trypsinogen, and chymotrypsinogen exist in an inactive or **proenzyme** state in the cell in which they are formed; they are subsequently activated by alteration of the proenzyme molecules. A study of the activation of a proenzyme should provide information about the active site of the resulting enzyme. One such proenzyme, trypsinogen, is synthesized by the pancreas and is converted to the active enzyme trypsin by the enzyme enterokinase or by trypsin itself. The primary structure of trypsinogen is a single polypeptide chain. The conversion to trypsin involves the splitting off of a hexapeptide, followed by a change in the conformation of the polypeptide to expose the catalytically active sites containing serine and histidine residues, which in the primary structure were far apart. This concept of the active site is represented in Figure 4–4.

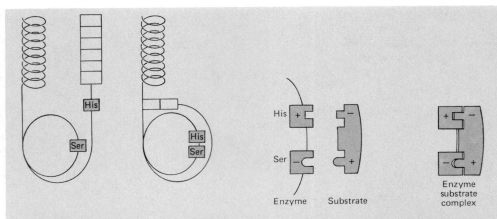

FIGURE 4-4 The conversion of trypsinogen to trypsin, showing the amino acids at the active site.

Another common method of identifying active sites is by selective inactivation of enzymes by reagents that react with known amino acids. Enzyme inhibitors such as diisopropylfluorophosphate and iodoacetate are used in these studies. Diisopropyl-fluorophosphate acts on certain enzymes to yield inactive derivatives in which the hydroxyl group of a specific serine residue is phosphorylated. In chymotrypsin, for example, this inhibitor selectively phosphorylates the serine residue at position 195, thus identifying this residue as essential at the active site. Several other enzymes, including trypsin, thrombin, elastase, and phosphorylase, have essential serine residues at their active sites and are also inhibited by diisopropylfluorophosphate. When the enzyme ribonuclease (p. 22) is treated with iodoacetate, the enzyme is alkylated and loses its catalytic activity. Two different inactive forms of the enzyme have been identified, one in which the imidazole ring of histidine residue 12 is alkylated, the other with alkylation of histidine 119. No other functional groups of ribonuclease were alkylated under these conditions at pH 5.5; therefore, it was concluded that histidine residues 12 and 119 are at the active site and are essential for catalytic activity.

To determine if other specific amino acid residues are involved in the active site of, for example, chymotrypsin near serine residue 195, the process of **affinity labeling** may be used. The enzyme is allowed to react with a synthetic molecule that resembles the true substrate in that it binds to serine 195 but in addition contains a functional group capable of reacting with some specific group of the enzyme on or near the active site. An affinity labeling agent was added to chymotrypsin which bound serine 195 but also possessed a potent alkylating group. This group reacted with histidine 57, indicating that both serine 195 and histidine 57 are at the active site of chymotrypsin.

Effect of Enzyme

When a purified enzyme is used, the rate of reaction is proportional to the concentration of the enzyme over a fairly wide range (Fig. 4-5). The substrate concentration must be kept constant and remain in excess of that required to combine with the enzyme. This relationship may also be used to measure the amount of an enzyme in a tissue extract or a biologic fluid. At the proper substrate concentration ($100 \times K_m$), temperature, and pH, the measured rate of activity is proportional to the quantity of enzyme present.

Effect of pH

The hydrogen ion concentration, or pH, of the reaction mixture exerts a definite influence on the rate of enzyme activity. If a curve is plotted comparing changes in pH

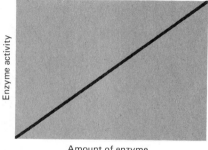

FIGURE 4-5   The effect of increasing amounts of enzyme on the activity of the enzyme.

with the rate of enzyme activity, it takes the form of an inverted U or V (Fig. 4–6). The maximum rate occurs at the **optimum pH,** with a rapid decrease of activity on either side of this pH value. The optimum pH of an enzyme may be related to a certain electric charge on the surface or to optimum conditions for the binding of the enzyme to its substrate. Most enzymes exhibit an optimum pH value close to 7, although pepsin is most active at pH 1.6 and trypsin at pH 8.2. Pepsin has no activity in an alkaline solution, whereas trypsin is inactive in an acid solution.

FIGURE 4-6   The effect of pH on enzyme activity.

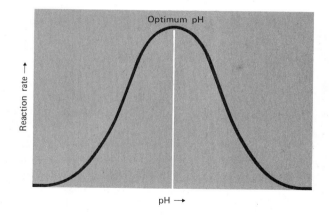

Effect of Temperature

The speed of most chemical reactions is increased two or three times for each $10°C$ rise in temperature. This is also true for reactions in which an enzyme is the catalyst, although the temperature range is fairly narrow. The activity range for most enzymes occurs between $10°$ and $50°C$; the **optimum temperature** for enzymes in the body is around $37°C$. The increased rate of activity observed at $50°C$ or above is short-lived, because the increased temperature first denatures and then coagulates the enzyme protein, thereby destroying its activity. The optimum temperature of an enzyme is therefore dependent on a balance between the rise in activity with increased temperature and the denaturation or inactivation by heat (Fig. 4–7). For any $10°$ rise in temperature the change in rate of enzyme activity is known as the Q_{10} value, or temperature coefficient. The Q_{10} value for most enzymes varies from 1.5 to 3.0.

Effect of End Products

The end products of an enzyme reaction have a definite effect on the rate of activity of the enzyme. If they are allowed to increase in concentration without removal, they

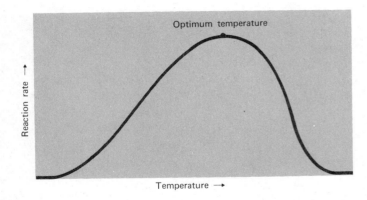

FIGURE 4-7 The effect of temperature on enzyme activity.

will slow the reaction. Some end products, when acid or alkaline in nature, may affect the pH of the mixture and thus decrease the rate of reaction. The effect of the end products on the activity of the enzyme is sometimes expressed as a chemical feedback system, with inhibition or decrease in rate called **negative feedback.** The activity of enzymes in the cell may be controlled to some extent by this chemical feedback system as in a sequence of cellular reactions in which the product D inhibits the enzyme reaction A → B.

$$A \xleftarrow{\hspace{0.3cm}} B \longrightarrow C \longrightarrow D$$

Several enzymes contain sulfhydryl groups (SH) which are associated with their active centers. Oxidizing agents change these groups to disulfide linkages and cause inactivation of the enzyme, whereas reducing agents restore the SH groups and activate the enzymes.

In the body many enzymes are secreted in an inactive form to prevent their action on the very glands and tissues that produce them. A proenzyme is the precursor of the active enzyme in the body. For example, pepsinogen is the proenzyme of pepsin and trypsinogen is the inactive form of trypsin. When pepsinogen is secreted into the stomach, it is converted into pepsin by hydrogen ions of the hydrochloric acid. The pepsin then activates more pepsinogen to form more of the active enzyme. Trypsinogen is secreted by the pancreas and is activated in the intestine by enterokinase (see p. 47 and Fig. 4–4). Also, during the process of purification an enzyme may become inactive. An enzyme may be activated by several agents: a change of pH, the addition of inorganic ions, or the addition of organic compounds.

The requirements for enzyme activity are further complicated by the fact that several enzymes require the presence of a metal ion for their activity. Enzymes have been characterized that require zinc, magnesium, iron, cobalt, and copper. Carbonic anhydrase, an enzyme that catalyzes the formation of carbonic acid from CO_2 and H_2O, requires zinc and is inactivated when this metal is removed.

ENZYME INHIBITORS

The activity of an enzyme may be inhibited by an increase in temperature, a change in pH, and the addition of a variety of protein precipitants. More specific inhibition can be achieved by the addition of an oxidizing agent to attack SH groups, or inhibitors such as iodoacetamide and *p*-chloromercuribenzoate that react with SH groups. Cyanide forms compounds with metals essential for enzyme action, whereas fluoride combines with magnesium and inhibits enzymes that require magnesium. Cyanide, for example, may remove a metal such as copper that is essential for the activity of the enzyme.

Sodium azide and monoiodoacetate are also potent inhibitors. This type of compound usually combines with a group at the active site of the enzyme and cannot be displaced by additional substrate. These inhibitors are called **noncompetitive inhibitors,** since their degree of inhibition is not related to the concentration of the substrate.

Compounds that directly compete with the substrate for the active site on the enzyme surface in the formation of the enzyme-substrate complex are called **competitive inhibitors.** An example of competitive inhibition would be the action of sulfanilamide on the utilization of *p*-aminobenzoic acid in the body. The similarity of these two compounds may readily confuse the enzyme involved in the utilization of this B vitamin in the synthesis of tetrahydrofolic acid, the active coenzyme of folic acid.

p-Aminobenzoic acid   Sulfanilamide

The enzyme succinic dehydrogenase catalyzes the oxidation of succinic acid to fumaric acid. Malonic acid inhibits this reaction, but the degree of inhibition can be reduced by the addition of more substrate, succinic acid. Malonic acid is similar to but has one less carbon atom than succinic acid as shown:

Malonic acid   Succinic acid

Malonic acid probably fits into the active site of succinic dehydrogenase, and thus it classifies as a competitive inhibitor. The concentration of the substrate and the concentration of the inhibitor both govern the action of malonic acid.

The Lineweaver-Burk plot (see p. 47) is frequently used to illustrate competitive or noncompetitive inhibition. In the example of malonic acid as an inhibitor of succinic dehydrogenase, the malonic acid combines with the dehydrogenase, forming an inactive enzyme inhibitor (EI) complex. The rate of formation of the product (fumaric acid) depends solely on the concentration of ES.

As long as there is sufficient substrate present in the reaction, the same $v_{max.}$ will be obtained, but the presence of EI alters the K_m for the substrate. In contrast to this behavior, noncompetitive inhibitors combine irreversibly with the enzyme molecule:

$$E + I \rightarrow EI$$

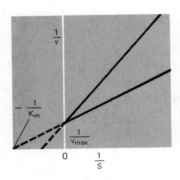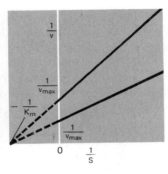

FIGURE 4–8 Lineweaver-Burk plot of competitive inhibition on the left compared to noncompetitive inhibition on the right.

As the inhibitor concentration is increased the amount of enzyme, E, is decreased and the $v_{max.}$ is proportionately decreased. The K_m value for the substrate remains constant, since $\frac{v_{max.}}{2}$ is proportional to $v_{max.}$. Competitive and noncompetitive inhibition are represented graphically in the Lineweaver-Burk plots shown in Figure 4–8.

The action of drugs in the body may depend on specific inhibitory effects on a particular enzyme in the tissues. The highly toxic nerve poison diisopropylfluorophosphate inhibits acetylcholine esterase, an enzyme essential for normal nerve function by forming an enzyme inhibitor compound by attachment to a hydroxyl group on a serine residue in the enzyme.

$$\text{Enzyme—OH} + \text{F—P}\overset{\displaystyle O}{\underset{}{\|}}\left(\text{O—C}\overset{\displaystyle CH_3}{\underset{H\ \ CH_3}{}}\right)_2 \rightarrow \text{Enzyme—O—P}\overset{\displaystyle O}{\underset{}{\|}}\left(\text{O—C}\overset{\displaystyle CH_3}{\underset{H\ \ CH_3}{}}\right)_2$$

Acetylcholine esterase	Diisopropyl-fluorophosphate (DFP)	Enzyme inhibitor compound

Antibiotic drugs may act by inhibiting enzyme and coenzyme reactions in microorganisms. Penicillin, for example, adversely affects cell wall construction in bacteria. A similar mechanism may be involved in the action of insecticides and herbicides.

Inhibitors of enzyme action in the body are called **antienzymes.** The tapeworm is a classic example of a protein-rich organism that is not digested in the intestine of the host. Substances that inhibit the activity of pepsin and trypsin have been isolated from the tapeworm. A trypsin inhibitor has been found in the secretion of the pancreas and in milk made from fresh soya beans. This substance exhibits properties similar to those of enzymes, and its activity is destroyed by heat. It may be formed by the pancreas to control the production of trypsin.

IMPORTANT TERMS AND CONCEPTS

active site
code number
competitive inhibitors
energy of activation
enzyme
Lineweaver-Burk plots

Michaelis constant
noncompetitive inhibitors
optimum pH
optimum temperature
specificity
substrate

QUESTIONS

1. What is the nature of an enzyme and how may it be defined?

2. What is a substrate? Give an example of a substrate and enzyme using the trivial name for the enzyme. Give an example of an enzyme and its substrate based on modern nomenclature.

3. Briefly explain how enzymes are classified.

4. How would you describe the activity of an enzyme with a systematic code number of 3.1.1.1.?

5. Give examples of three types of specificity of action of enzymes.

6. What is meant by the energy of activation of a reaction? How is it affected by an enzyme?

7. What role does the formation of an enzyme-substrate complex play in the action of an enzyme? Explain.

8. Draw a graph representing the change in activity of an enzyme as the amount of its substrate is increased from zero to a maximum concentration. Explain the shape of the curve obtained.

9. How would you define the Michaelis constant, K_m, of an enzyme? Why is the K_m value important in enzyme reactions?

10. What are the advantages of expressing an enzyme reaction as a Lineweaver-Burk plot?

11. What is meant by the active site of an enzyme? Explain.

12. What is meant by (1) optimum pH? (2) optimum temperature of enzyme reactions?

13. Cyanide is a very potent poison. Explain how cyanide may exert its toxic properties.

14. Explain the difference between competitive and noncompetitive inhibitors, and give an example of each.

15. Illustrate competitive and noncompetitive inhibition using Lineweaver-Burk plots.

Chapter 5

CARBOHYDRATES

The *objectives* of this chapter are to enable the student to:

1. Distinguish between the (+) and (−) forms of an optically active compound.
2. Describe the structure of α-D(+) glucose and its relation to α-D-glucopyranose.
3. Recognize the α and β isomers of the pyranose and furanose forms of sugars.
4. Illustrate and explain the difference between the Haworth and the chair form of α-D-glucopyranose.
5. Describe the glycoside or acetal linkage between two monosaccharides.
6. Write the formula for sucrose and explain why it does not reduce Benedict's solution.
7. Illustrate the structure of a polysaccharide using several molecules of glucose.
8. Recognize the difference between the structures of cellulose and glycogen.

Carbohydrates are less plentiful than proteins, making up about 10 per cent of the organic matter of the living cell. There are probably about 50 different kinds of carbohydrates in the cell. They play important physiological roles as a source of energy for the cell, as storage of chemical energy, as structural units in cell walls and membranes, and in cellular components responsible for function and growth.

As a class of compounds, carbohydrates include simple sugars, starches, and celluloses. Simple sugars such as glucose, fructose, and sucrose are constituents of many fruits and vegetables. Starches are the storage form of carbohydrates in plants. Cellulose is the main supporting structural material of trees and plants, and is also the base material for the manufacture of many polymers and plastics of commercial importance. About 75 per cent of the solid matter of plants consists of carbohydrates.

OPTICAL ACTIVITY

Stereoisomerism is a common phenomenon in organic chemistry. These isomers may be **structural, geometric,** or **optical.** Optical isomerism is frequently encountered in organic compounds of biochemical interest and is essential in a study of the composition, properties, and reactions of carbohydrates. Many organic molecules, including the carbohydrates, exhibit the phenomenon of **optical activity.** Any optically active compound possesses the property of rotating a plane of polarized light. Ordinary light may be thought of as radiant energy propagated in the form of wave motion whose vibrations take place in all directions at right angles to the path of the beam of light. If a light ray is considered as perpendicular to the plane of the page and passing through it, the vibrations may be

FIGURE 5-1   Rays of ordinary light coming toward the observer and vibrating in all planes at right angles to the path of the light, compared to the rays of plane polarized light vibrating in only one plane.

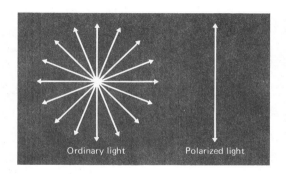

Ordinary light   Polarized light

represented as spokes on a wheel (Fig. 5-1). Certain minerals, such as Iceland spar and tourmaline, and polaroid sheets or discs (properly oriented crystals embedded in a transparent plastic) allow only light vibrating in a single plane to pass through their crystals. When ordinary light is passed through a Nicol prism, which consists of two pieces of tourmaline cemented together, the resulting beam is traveling in one direction and in one plane, and is called **plane polarized** or just **polarized** light (Fig. 5-1).

Carbohydrates in solution show the property of optical rotation, i.e., a beam of polarized light is rotated when it passes through the solution. The extent to which the beam is rotated, or the angle of rotation, is determined with an instrument called a **polarimeter.** A simple polarimeter is shown in Figure 5-2.

The two Nicol prisms are arranged in such a manner that the light readily passes through them to illuminate the eyepiece uniformly. A cylindrical cell containing a solution of an optically active substance is then placed between the two Nicol prisms. Since the solution rotates the plane polarized light, it will not pass through the second Nicol prism and the field of the eyepiece will become dark. The analyzing Nicol prism is then rotated until the field of the eyepiece is again uniformly illuminated, and the number of degrees through which the prism is rotated as well as the extent of rotation can be determined. A substance whose solution rotates the plane of polarized light to the right is said to be **dextrorotatory;** one whose solution rotates the light to the left is called **levorotatory.** The rotation is designated $(+)$ for dextrorotatory or $(-)$ for levorotatory; for example, $(+)$lactic acid and $(-)$lactic acid.

To standardize the experimental work with the polarimeter, the term **specific rotation** has been adopted. The specific rotation $[\alpha]_D^{20°}$ of a substance is the rotation in angular degrees produced by a column of solution 1 decimeter long, in which the concentration is 1 gram per ml. The terms 20° and D refer to the temperature of the solution, 20°C, and the source of plane polarized light, which is the monochromatic sodium light of wavelength 589.0 to 589.6 nm corresponding to the D line in the yellow part of the spectrum.

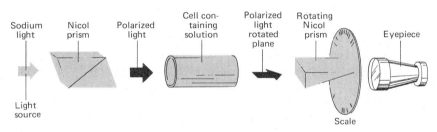

Sodium light   Nicol prism   Polarized light   Cell containing solution   Polarized light rotated plane   Rotating Nicol prism   Eyepiece

Light source

Scale

FIGURE 5-2   A diagrammatic sketch of the essential components of a polarimeter.

Van't Hoff and La Bel independently advanced the same theory to explain the fundamental reason for the optical activity of a compound. They postulated that the presence of an **asymmetric carbon atom** in a compound was responsible for the optical activity.[*]

Although the presence of an asymmetric carbon atom is most frequently responsible for the optical activity of a compound, any condition that removes the elements of symmetry that make mirror images identical produces optical activity. A molecule such as tartaric acid with two asymmetric carbon atoms that are alike (attached to the same four kinds of groups) may exist in a form in which there is a plane of symmetry and each half molecule is a mirror image of the other. This is known as the *meso* form and exhibits an optical activity of zero. Structures of tartaric acid representing the *meso* form are shown as follows:

$$
\begin{array}{ccc}
\overset{\displaystyle O}{\overset{\|}{C}}-OH & \qquad & \overset{\displaystyle O}{\overset{\|}{C}}-OH \\
H-C-OH & & HO-C-H \\
\cdots\cdots & & \cdots\cdots \\
H-C-OH & & HO-C-H \\
\overset{\displaystyle}{C}\overset{\displaystyle O}{\underset{\|}{}}-OH & & C-OH
\end{array}
$$

meso-Tartaric acid

A compound having two like asymmetric carbon atoms, therefore, has only three optical isomers, the (+) and (−), and the *meso* form. Before Van't Hoff and La Bel postulated the presence of asymmetric carbon atoms, Pasteur recognized two forms of crystals of tartaric acid and carefully separated them by hand. One form was dextrorotatory in solution, the other levorotatory.

Lactic acid from different natural sources exhibits differences in optical activity. For example, the lactic acid involved in the contraction of muscle tissue in the body is the dextro form, whereas the levo form may be isolated from the fermentation products of cane sugar. When milk sours, the lactic acid that is formed consists of an equal mixture of the (+) and (−) forms and does not rotate the plane of polarized light. In general, when a compound that exhibits optical activity is synthesized in the laboratory, a mixture of equal parts of the (+) and (−) forms results. Such a mixture is called a **racemic mixture.** Reactions carried out in the body or in the presence of microorganisms often produce optically active isomers, since the reactions are catalyzed by enzymes which themselves are optically active. The enzyme reactions are often specific for the (+) or (−) component of a racemic mixture.

As naturally occurring optically active compounds such as the carbohydrates and the amino acids are studied, it will be seen that the isomeric form, and therefore the structure, is closely related to the physiological activity of these substances.

COMPOSITION

The classic definition of carbohydrates stated that they were compounds of C, H, and O in which the H and O were in the same proportion as in water. But a compound such as acetic acid, $C_2H_4O_2$, fits this definition and yet is not classed as a carbohydrate, whereas an important carbohydrate such as deoxyribose, $C_5H_{10}O_4$, a constituent of DNA (deoxyribonucleic acid) found in every cell, does not fit the definition. Carbohydrates are now defined as derivatives of polyhydroxyaldehydes or polyhydroxyketones. A sugar that contains an aldehyde group is called an **aldose,** and one that contains a ketone group is termed a **ketose.**

---

[*]An asymmetric carbon is one that has four different groups attached to it.

CLASSIFICATION

The simplest carbohydrates are known as **monosaccharides,** or simple sugars. Monosaccharides are derivatives of straight-chain polyhydric alcohols and are classified according to the number of carbon atoms in the chain. A sugar with two carbon atoms is called a diose; with three, a triose; with four, a tetrose; with five, a pentose; and with six, a hexose. The ending -ose is characteristic of sugars. When two monosaccharides are linked together by splitting out a molecule of water, the resulting compound is called a **disaccharide.** The combination of three monosaccharides results in a **trisaccharide,** although the general term for carbohydrates composed of two to five monosaccharides is **oligosaccharide.** Polymers composed of several monosaccharides are called **polysaccharides.**

Carbohydrates which will be considered in this chapter may be classified as follows:

I. Monosaccharides
  Trioses—$C_3H_6O_3$
    Aldose—Glyceraldehyde
    Ketose—Dihydroxyacetone
  Pentoses—$C_5H_{10}O_5$
    Aldoses—Arabinose
      Xylose
      Ribose
  Hexoses—$C_6H_{12}O_6$
    Aldoses—Glucose
      Galactose
    Ketoses—Fructose
  Ascorbic acid

II. Disaccharides—$C_{12}H_{22}O_{12}$
  Sucrose (glucose + fructose)
  Maltose (glucose + glucose)
  Lactose (glucose + galactose)

III. Polysaccharides
  Hexosans
    Glucosans—Starch
      Glycogen
      Dextrin
      Cellulose

IV. Mucopolysaccharides
  Hyaluronic acid
  Chondroitin sulfate
  Heparin

Trioses

The trioses are important compounds in muscle metabolism, and are the basic sugars to which all monosaccharides are referred. The definition of a simple sugar may readily be illustrated by the use of the trioses. The polyhydric alcohol from which they are derived is glycerol. Oxidation on the end carbon atom produces the aldose sugar known as glyceraldehyde; oxidation on the center carbon yields the keto triose, dihydroxyacetone. It can be seen from the formula of glyceraldehyde that one asymmetric carbon atom

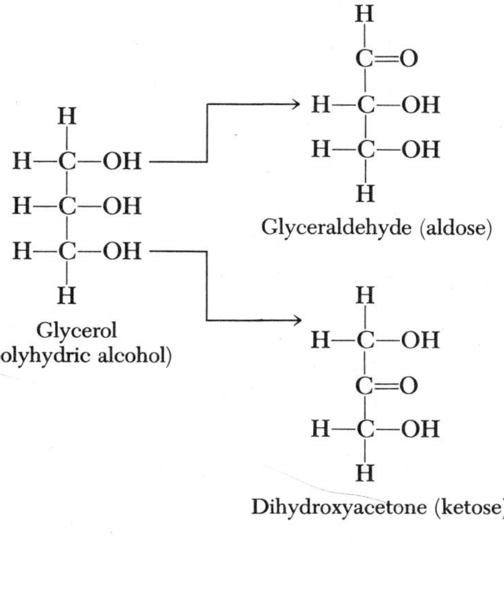

Glyceraldehyde (aldose)

Glycerol
(polyhydric alcohol)

Dihydroxyacetone (ketose)

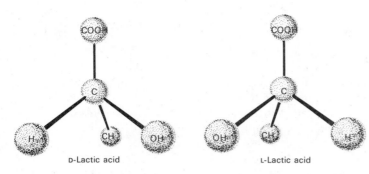

FIGURE 5–3   The spatial relationship of the groups attached to the asymmetric carbon atom of the D and L forms of lactic acid.

is present. Therefore this sugar can exist in two forms, one of which rotates plane polarized light to the right, the other to the left. Modern terminology employs the D and L, written in small capital letters, for structural relationships, and a (+) and (−) for direction of rotation of polarized light.

D (+) Glyceraldehyde   L (−) Glyceraldehyde   Perspective formula

Fischer projection formula

The isomeric forms of sugars are often represented as the **Fischer projection formula.** The asymmetric carbon atom * of glyceraldehyde would represent the central sphere as in the model in Figure 5–3, with the H and OH groups projecting in front of the plane of the paper and the aldehyde and primary alcohol group projecting behind. As ordinarily written, the horizontal bonds are understood to be in front of the plane and the vertical bonds behind the plane of the paper. The **perspective formula** emphasizes the position of the groups using dotted lines to connect those behind the plane and heavy wedges to represent groups in front of the plane. The Fischer projection formulas are always written with the aldehyde or ketone groups (the most highly oxidized groups) at the top of the structure; therefore, all monosaccharides with the hydroxyl group on the right of the carbon atom next to the bottom primary alcohol group are related to D-glyceraldehyde and are called D-sugars.

In like manner, if the hydroxyl group on the carbon atom next to the end primary alcohol group is on the left, it is related to L-glyceraldehyde and is an L-sugar. The direction of rotation of polarized light cannot be ascertained from the formula, but must be determined experimentally.

Only two optical isomers of aldotriose exist, since it contains only one asymmetric carbon atom. A sugar such as a tetrose with two structurally different carbon atoms would exhibit a total of four different isomers, and as each new asymmetric carbon atom is added to the structure the number of isomers is doubled. The maximum number of isomers of a sugar is 2^n, where n is the number of different asymmetric carbon atoms. For example, pentoses contain three asymmetric carbon atoms and can form eight isomers; hexoses contain four asymmetric carbon atoms and can exist as 16 different isomers.

Pentoses

The pentoses are sugars whose molecules contain five carbon atoms and three asymmetric carbon atoms. They occur in nature combined in polysaccharides from which the monosaccharides may be obtained by hydrolysis with acids. Arabinose is obtained from gum arabic and the gum of the cherry tree, and xylose is obtained by hydrolysis of wood, corn cobs, or straw. Ribose and deoxyribose are constituents of the ribose nucleic acids, RNA, and deoxyribose nucleic acids, DNA, that are essential components of the cytoplasm and nuclei of cells.

Hexoses

The hexoses are by far the most important monosaccharides from a nutritional and physiological standpoint. The bulk of the carbohydrates used as foods consist of hexoses free or combined in disaccharides and polysaccharides. Glucose, fructose, and galactose are the hexoses commonly occurring in foods, whereas mannose is a constituent of a vegetable polysaccharide. Glucose, also called **dextrose,** is the normal sugar of the blood and tissue fluids and is utilized by the cells as a source of energy. Fructose often occurs free in fruits and is the sweetest sugar of all the monosaccharides. Galactose is a constituent of **milk sugar** and is found in brain and nervous tissue. All these monosaccharides are D-sugars. The hydroxyl group on the carbon next to the primary alcohol is on the right. Fructose is a ketose sugar and the others are aldose sugars.

D-Glucose
(aldose)

D-Fructose
(ketose)

D-Galactose
(aldose)

Although glucose and galactose are represented as simple aldehyde structures, this form does not explain all the reactions they undergo. Both of these aldoses, for example, do not give a characteristic test for aldehyde. Also, when a glucose solution is allowed to stand, a change in its specific rotation may be observed.

Freshly prepared aqueous solutions of crystalline glucose often yield a specific rotation as high as +113 degrees, whereas glucose crystallized from pyridine exhibits a specific rotation as low as +19 degrees. On standing, both of these solutions change their rotation until an equilibrium value of +52.5 degrees is reached. This change in rotation is called **mutarotation.** Since the specific rotation of an organic compound is related to its structure, as is its melting point, boiling point, and other properties, it may

α-D-Glucose
+113°

D-Glucose,
straight chain
form

β-D-Glucose
+19°

be suspected that glucose exists in two different isomeric forms. This has been shown to be true and is explained by the existence of an **intramolecular bridge structure** involving carbon atoms 1 and 5. It can be seen from these formulas that the free aldehyde group no longer exists and a new asymmetric carbon is produced. To indicate the position of the hydroxyl group on the first carbon and to distinguish between the two new isomers, the α-isomer has the OH on the right and the β-isomer on the left as shown. When the α- or β-isomer is dissolved in water, an equilibrium mixture of 37 per cent α and 63 per cent β, with a specific rotation of 52.5 degrees, is formed. This intramolecular bridge structure and the phenomenon of mutarotation are common to all aldohexoses, and since the structure contains an additional asymmetric carbon atom, the number of possible isomers is doubled.

A further projection of the structure results from the random motion of the open chain which allows the alcoholic hydroxyl group on carbon-5 of the aldohexose molecule to approach the aldehyde group on the end of the molecule. Ring formation could then result from the formation of a hemiacetal between the aldehyde and hydroxyl group. The hemiacetal formation in an aldohexose molecule may be represented as follows:

Haworth suggested that the sugars be represented as derivatives of the heterocyclic rings pyran and furan.

Pyran   Furan

The relation between the straight chain structure of glucose and Haworth's **glucopyranose** may best be understood by writing glucose in a chain as shown in A. The chain is folded and a rotation of groups occurs around carbon-5 to bring the primary alcohol group (carbon-6) into the proper spatial relation to the other groups (B). The hemiacetal is then formed between the aldehyde on carbon-1 and the OH group on carbon-5 to form α-D-glucopyranose, shown in C.

Glucose α-D-Glucopyranose

A   B   C

The heavy lines represent the base of a space model in which the five carbons and one oxygen are in the same plane perpendicular to the plane of the paper. The thick bonds of the ring extend toward the reader, whereas the thin bonds of the ring are behind the plane of the paper. Groups which are ordinarily written on the right of the oxide ring structure appear below the plane of the pyranose ring, and those to the left of the carbon chain appear above the plane. The glucopyranose structure, C, has the OH on carbon-1 below the plane of the ring and would therefore be an α-isomer.

Monosaccharides such as the pentoses and fructose, whose oxide rings enclose four carbon atoms, are written as derivatives of furan as shown:

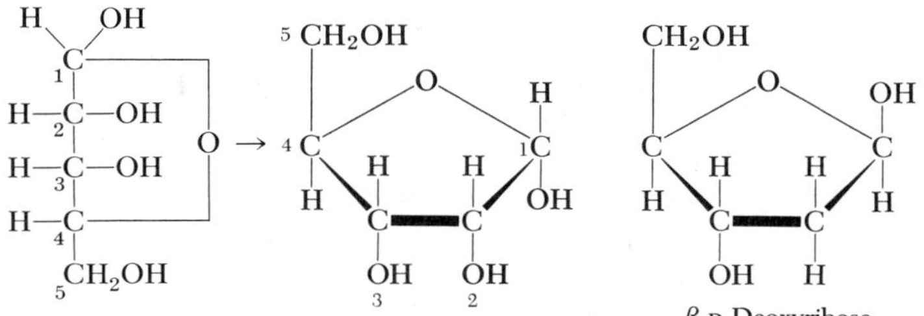

α-D-Ribose

β-D-Deoxyribose

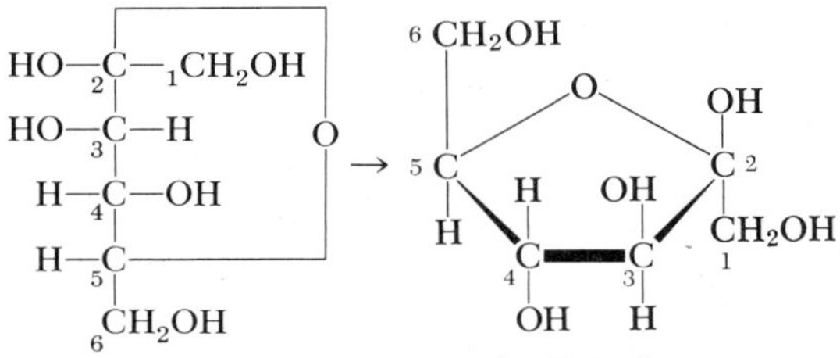

β-D-Fructofuranose

It can readily be observed in the preceding structures that the OH group on carbon-1 (carbon-2 in ketofuranoses) indicates the α or β form of the sugar. The α-isomers have the OH extending below the plane of the ring, whereas in the β-isomers the OH extends above the plane.

The Haworth structural projections may be somewhat misleading, since they suggest that the five- and six-membered furanose and pyranose rings are planar, and this is not the case. The pyranose ring exists in two conformations, the chair and the boat form. The chair form is more stable than the boat form, and is the predominant form in aqueous solutions of hexoses. The stable conformation (chair form) of α-D-glucopyranose is usually represented with the edge of the ring nearest the reader in bold lines, as in the Haworth structure.

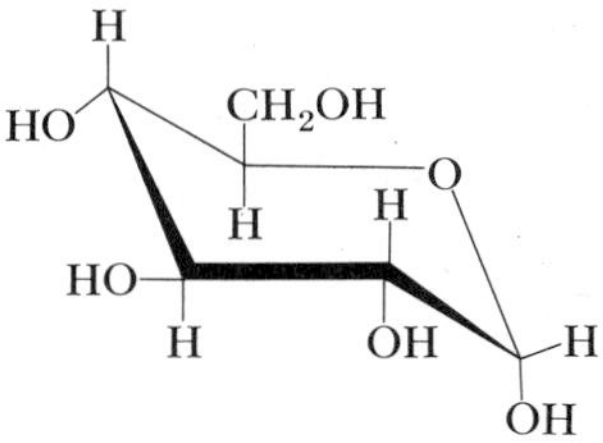

α-D-Glucopyranose

REACTIONS OF CARBOHYDRATES

Dehydration

When aldohexoses or aldopentoses are heated with strong acids, they are dehydrated to form furfural derivatives. Pentoses yield furfural whereas aldohexoses form hydroxymethyl furfural.

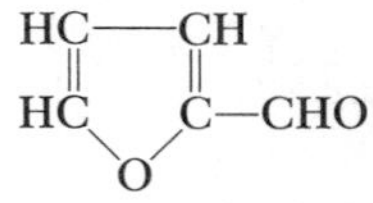

Furfural 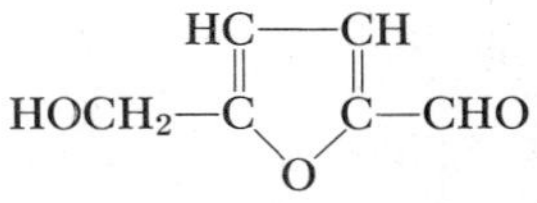   Hydroxymethyl furfural

The furfural derivatives formed in this reaction combine with α-naphthol to give a purple color. This color is the basis of the Molisch test, a general test for carbohydrate. Furfural reacts with orcinol to yield a green color which constitutes Bial's test for pentoses.

Acetal or Glycoside Formation

When monosaccharides are treated with an alcohol in a strong acid solution, they form **glycosides.** The hemiacetal structure reacts with alcohols or an alcoholic hydroxyl group to form an acetal or glycoside.

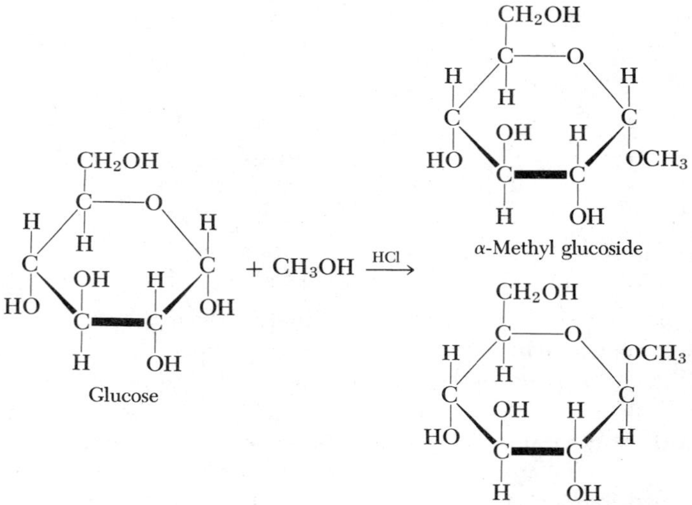

α-Methyl glucoside

β-Methyl glucoside

Glucose

The position of the methyl group below or above the plane of the ring indicates α- or β-methyl glucoside in that order. This is a very important reaction since many of the disaccharides and polysaccharides are glycosides in which one of the alcoholic hydroxyl groups in the second monosaccharide reacts with the hemiacetal in the first, as shown below:

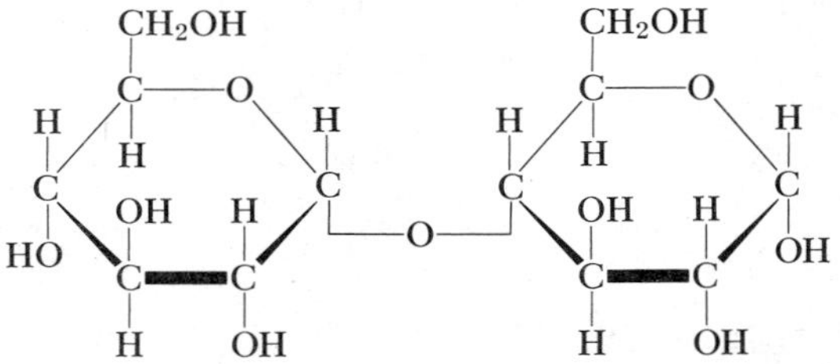

α-1-4 Linked disaccharide

When glucose, fructose, or mannose is added to a saturated solution of $Ba(OH)_2$ and allowed to stand, it forms the same intermediate **enediol.** The loss of asymmetry in the second carbon atom of the intermediate enediol favors the formation of the other two sugars. The alkaline solution favors the formation of enediols and suppresses the formation of ring structures. This is called the **Lobry de Bruyn-von Eckenstein transformation** and may be illustrated as shown:

$$
\begin{array}{l}
\boxed{\begin{array}{c} H\text{—}C\text{=}O \\ HO\text{—}C\text{—}H \end{array}} \\
HO\text{—}C\text{—}H \\
H\text{—}C\text{—}OH \quad \text{Mannose} \\
H\text{—}C\text{—}OH \\
CH_2OH
\end{array}
$$

$$\updownarrow$$

$$
\boxed{\begin{array}{c} H\text{—}C\text{=}O \\ H\text{—}C\text{—}OH \end{array}}
\rightleftharpoons
\boxed{\begin{array}{c} H\text{—}C\text{—}OH \\ C\text{—}OH \end{array}}
\rightleftharpoons
\boxed{\begin{array}{c} H_2\text{—}C\text{—}OH \\ C\text{=}O \end{array}}
$$

HO—C—H	HO—C—H	HO—C—H
H—C—OH	H—C—OH	H—C—OH
H—C—OH	H—C—OH	H—C—OH
CH₂OH	CH₂OH	CH₂OH
Glucose	Enediol	Fructose

Oxidation

One of the important reactions of carbonyl compounds is the oxidation to carboxylic acids. Sugars that contain *free or potential aldehyde or ketone groups* in the hemiacetal type structure are oxidized in alkaline solution by Cu^{+2} and Ag^+. This reaction is the basis of the Benedict's, Fehling's, and silver mirror tests. Sugars that undergo oxidation in these reactions are called **reducing sugars.** The determination of reducing sugars is often used for the estimation of glucose in the blood and urine. More recently, the enzyme **glucose oxidase** has been used in these determinations because of its greater specificity.

All the reducing sugars that are capable of reducing Cu^{+2} to Cu^+ are oxidized in the reaction described in the preceding section. If the aldehyde group of glucose is oxidized to a carboxyl group by a weak oxidizing agent, such as NaOBr, gluconic acid is formed. Oxidation of the primary alcohol group, either by chemical agents or enzymes, produces glucuronic acid. Further oxidation with concentrated HNO_3 converts both end groups into carboxyl groups as in saccharic acid.

COOH	H C=O	COOH
H—C—OH	H—C—OH	H—C—OH
HO—C—H	HO—C—H	HO—C—H
H—C—OH	H—C—OH	H—C—OH
H—C—OH	H—C—OH	H—C—OH
CH₂OH	COOH	COOH
Gluconic acid	Glucuronic acid	Saccharic acid

Glucuronic acid combines with drugs and toxic compounds in the body, and the conjugated glucuronides are excreted in the urine. Oxidation of galactose with concentrated

HNO_3 produces mucic acid. This acid crystallizes readily and the reaction is used as a test for the presence of galactose.

Fermentation

The enzyme mixture called **zymase** present in common bread yeast will act on some of the hexose sugars to produce alcohol and carbon dioxide. The fermentation of glucose may be represented as follows:

$$C_6H_{12}O_6 \xrightarrow{\text{zymase}} 2C_2H_5OH + 2CO_2$$

Glucose   Ethyl alcohol

The common hexoses (with the exception of galactose) ferment readily, but pentoses are not fermented by yeast. Disaccharides must first be converted into their monosaccharide constituents by other enzymes present in yeast before they are susceptible to fermentation by zymase.

There are many other types of fermentation of carbohydrates besides the common alcoholic fermentation. When milk sours, the lactose of milk is converted into lactic acid by a fermentation process. Citric acid, acetic acid, butyric acid, and oxalic acid may all be produced by special fermentation processes.

Ester Formation

Esters formed between the hydrogen atom of a hydroxyl group of phosphoric acid and the hydroxyl group of a monosaccharide are common, and several of these phosphorylated sugars are encountered in carbohydrate metabolism (Chapter 10). In the metabolic reactions in the body the location of the hydroxyl group or groups on the compounds to be phosphorylated is controlled by enzymes specific for that reaction. The organic chemist utilizes phosphoryl group donors in nonaqueous systems and chemically blocks other potentially reactive groups in order to phosphorylate the desired hydroxyl group in a compound.

α-D-Glucose-6-phosphate   α-D-Fructose-1, 6-diphosphate

DISACCHARIDES

A disaccharide is composed of two monosaccharides whose combination involves the splitting out of a molecule of water. The acetal linkage is always made from the aldehyde group of one of the sugars to a hydroxyl or ketone group of the second. In the structure of the individual disaccharides, such as sucrose, lactose, and maltose, the exact location of the acetal linkage and the isomeric forms involved in the linkage are obviously important in the proof of the structure of the disaccharide. Information concerning the monosaccharides formed by hydrolysis, the reduction products, the change in optical rotation on hydrolysis, and the reaction as a reducing sugar is used in the proof of the structure. In order to reduce Benedict's solution, disaccharides must have a potential aldehyde or ketone group that is not involved in the acetal linkage between the two sugars.

Sucrose

Sucrose is commonly called **cane sugar** and is the ordinary sugar that is used for sweetening purposes in the home. It is found in many plants such as sugar beets, sorghum cane, the sap of the sugar maple, and sugar cane. Commercially it is prepared from sugar cane and sugar beets.

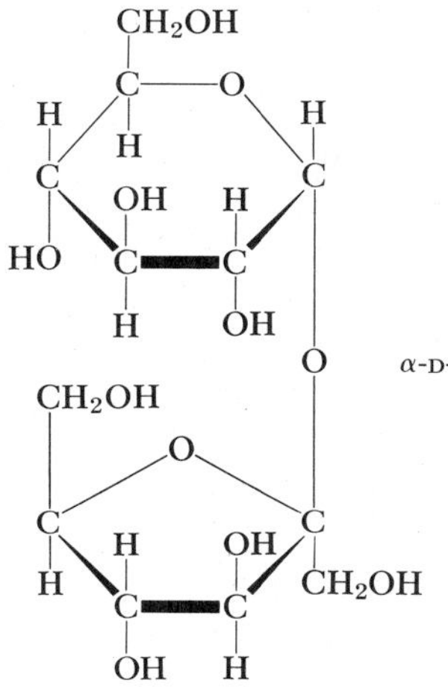

α-D-Glucopyranosyl-β-D-fructofuranose

Sucrose

Sucrose is composed of a molecule of glucose joined to a molecule of fructose in such a way that the linkage involves the reducing groups of both sugars (carbon-1 of glucose and carbon-2 of fructose). It is the only common mono- or disaccharide that will not reduce Benedict's solution. When sucrose is hydrolyzed, either by the enzyme sucrase or by an acid, a molecule of glucose and a molecule of fructose are formed. The fermentation of sucrose by yeast is possible, since the yeast contains the two enzymes sucrase and zymase. The sucrase first hydrolyzes the sugar, and then the zymase ferments the monosaccharides to form alcohol and carbon dioxide.

Lactose

The disaccharide present in milk is lactose, or **milk sugar.** It is synthesized in the mammary glands of animals from the glucose in the blood. Commercially, it is obtained from milk whey and is used in infant foods and special diets. Lactose, when hydrolyzed by the enzyme lactase or by an acid, forms a molecule of glucose and a molecule of

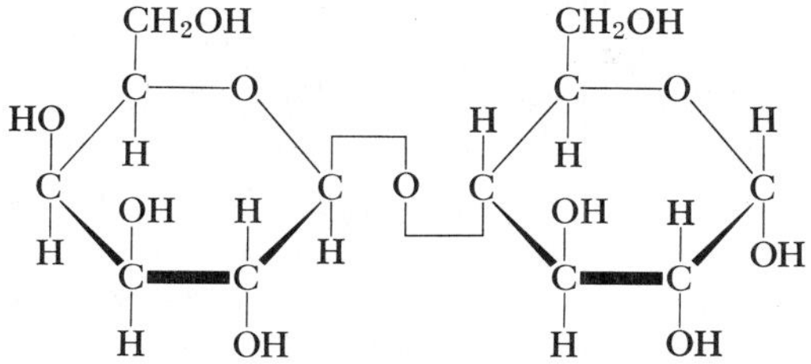

β-D-Galactopyranosyl-α-D-glucopyranose

Lactose

galactose. Lactose will reduce Benedict's solution, but is not fermented by yeast. From its reducing properties, it is obvious that the linkage between its constituent monosaccharides does not involve both potential aldehyde groups (carbon-1 of galactose is connected to carbon-4 of glucose). This linkage also illustrates the formation of an ether linkage by dehydration.

Maltose

Maltose is present in germinating grains. Since it is obtained as a product of the hydrolysis of starch by enzymes present in malt, it is often called **malt sugar.** It is also formed in the animal body by the action of enzymes on starch in the process of digestion. Commercially, it is made by the partial hydrolysis of starch by acid in the manufacture of corn syrup. Maltose reduces Benedict's solution and is fermented by yeast. On hydrolysis it forms two molecules of glucose.

α-D-Glucopyranosyl-α-D-glucopyranose
Maltose

Cellobiose

Cellobiose is similar to maltose since it is composed of two glucose molecules joined in a carbon-1 to carbon-4 linkage. In contrast to maltose, the glucose in cellobiose is in the β form.

Cellobiose

POLYSACCHARIDES

The polysaccharides are complex carbohydrates that are made up of many monosaccharide molecules and therefore possess a high molecular weight. They differ from the simple sugars in many ways. They do not have a sweet taste, are usually insoluble in water, and when dissolved by chemical means form colloidal solutions because of their large molecules. Although most polysaccharides have a terminal monomer present as a reducing sugar, the contribution of this portion to the properties of the molecule decreases as the size of the polymer increases. Most polysaccharides, therefore, do not behave as reducing sugars, although most oligosaccharides will reduce Benedict's reagent.

There are polysaccharides formed from pentoses or from hexoses, and there are also mixed polysaccharides. Of these, the most important are composed of the hexose glucose and are called **hexosans,** or more specifically, **glucosans.** As in a disaccharide, whenever two molecules of a hexose combine, a molecule of water is split out. For this reason, a hexose polysaccharide may be represented by the formula $(C_6H_{10}O_5)_x$. The x represents the number of hexose molecules in the individual polysaccharide. Because of the complexity of the molecules, the number of glucose units in any one polysaccharide is still an estimate. In addition, the molecular weight values obtained for polysaccharides from different plant and animal sources show considerable variation. For this reason the molecular weights of the polysaccharides described in the following section are only approximations.

Starch

From a nutritional standpoint, starch is the most important polysaccharide. It is made up of glucose units and is the storage form of carbohydrates in plants. It consists of two types of polysaccharides: **amylose,** composed of a chain of glucose molecules connected

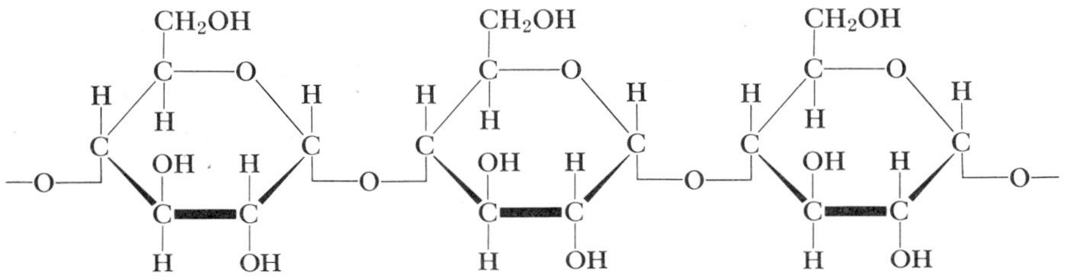

Amylose

by α-1,4 linkages, and **amylopectin,** which is a branched chain or polymer of glucose with both α-1,4 and α-1,6 linkages. The repeating structure of glucose molecules in amylose is usually represented as glucopyranose units as shown in the accompanying diagram. Amylose chains vary in molecular weight from a few thousand to 500,000, compared to molecular weights up to 100 million for amylopectin. The branching of the glucose chain in amylopectin occurs about every 24 to 30 glucose molecules.

Starch will not reduce Benedict's solution and is not fermented by yeast. When starch is hydrolyzed by enzymes or by an acid, it is split into a series of intermediate compounds possessing small numbers of glucose units. The product of complete hydrolysis is the free glucose molecule. A characteristic reaction of starch is the formation of a blue compound with iodine. This test is often used to follow the hydrolysis of starch, since the color changes from blue through red to colorless with decreasing molecular weight:

starch → amylodextrin → erythrodextrin → achroodextrin → maltose → glucose
  blue   blue   red   colorless   colorless with iodine

Dextrins

Dextrins are found in germinating grains, but are usually obtained by the partial hydrolysis of starch. Those formed from amylose have straight chains, whereas those derived from amylopectin exhibit branched chains of glucose molecules. The larger branched chain molecules give a red color with iodine and are the erythrodextrins. They are soluble in water and have a slightly sweet taste. Large quantities of dextrins are used in the manufacture of adhesives because they form sticky solutions when wet. An example of their use is the mucilage on the back of postage stamps.

Glycogen

Glycogen is the storage form of carbohydrate in the animal body and is often called animal starch. It is found in liver and muscle tissue, is soluble in water, does not reduce

Benedict's solution, and gives a red-purple color with iodine. The glycogen molecule is similar to the amylopectin molecule in that it has branched chains of glucose with α-1,4 and α-1,6 linkages that occur about every 4 to 5 glucose molecules. The branched

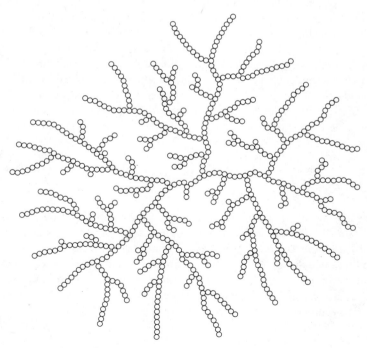

Branched chain of glucose molecules in amylopectin and glycogen

chain structure common to both glycogen and amylopectin is shown above and represented in Figure 5–4.

The molecular weight of glycogen is large and often exceeds 5,000,000. When glycogen is hydrolyzed in the animal body, it forms glucose to help maintain the normal sugar content of the blood.

FIGURE 5-4   Representation of the branched chain structure of glycogen. (Adapted from McGilvery: Biochemical Concepts, Philadelphia, W. B. Saunders Company, 1975, p. 295.)

Cellulose

Cellulose is a polysaccharide that occurs in the framework, or supporting structure, of plants. It is composed of a straight chain polymer of glucose molecules similar in structure to that pictured for amylose. The major difference concerns the linkage of the glucose molecules. In amylose the linkage is α-1,4, as in maltose, whereas in cellulose the linkage is β-1,4, which occurs in **cellobiose.** The maltose type of structure is hydrolyzed by enzymes and serves as a source of dietary carbohydrate. In contrast, the

Cellulose

cellobiose structure is insoluble in water, will not reduce Benedict's solution and is not attacked by enzymes present in the human digestive tract. Ruminants, such as the cow, can use cellulose for food since bacteria in the rumen form the enzyme **cellulase,** which hydrolyzes cellulose to glucose. The molecular weight of cellulose in different species of plants has been estimated to range from 50,000 to 2,500,000, which is equivalent to 300 to 15,000 glucose residues.

TOPIC OF CURRENT INTEREST

THE MANY FACES OF CELLULOSE

Electron microscope studies show that in the cell walls of plants, densely packed cellulose fibrils surround the cell in regular parallel arrangements, often in criss-cross layers. These fibrils are cemented together by a matrix of other natural polymers such as pectin and hemicellulose. The cell walls of higher plants have often been compared to reinforced concrete, in which the cellulose fibrils correspond to the steel rods and the material in the matrix to the concrete. Because of this reinforcement, plant walls are capable of withstanding the large osmotic pressure difference between the extracellular and intracellular fluid compartments without swelling of the cell. Their structure contributes physical strength and rigidity to the stems, leaves, and root tissues of plants and allows them to carry large weights and resist physical stress.

Cotton, which is almost pure cellulose, grows like hair on the seeds of the cotton plant. When the plant is mature, each hair or fiber has developed into a bundle of cellulose molecules lying side by side in a special arrangement. These fibers are readily spun into thread, which may be woven into cotton cloth. For many years articles of clothing made of cotton were less espensive and thought less stylish than those made of silk, nylon, Dacron, and polyester. At present, however, cotton is enjoying a renewed popularity because its cloth is comfortable, has the ability to "breathe," and absorbs moisture and accepts dyestuffs more readily than the other fibers. When cotton thread or cloth is treated under tension with a concentrated solution of sodium hydroxide, it takes on a silk-like luster and increases in strength. The product of this process is called **mercerized cotton,** which is used in large quantities in the manufacture of cotton cloth.

Many important polymers are made from cellulose. The artificial fiber **rayon** was manufactured as early as 1911. In making rayon, cellulose in the form of cotton is dissolved by treatment with sodium hydroxide, followed by carbon disulfide. The alkaline solution of cellulose is then forced through fine holes into dilute sulfuric acid to make the rayon fibers. A similar process is used to produce cellophane, where sheets of polymer rather than fibers are made. When treated with nitric acid, cellulose is converted into cellulose nitrates, which are esters of commercial importance. Guncotton is a nitrocellulose containing about 13 per cent nitrogen; it is used in the production of smokeless powder and high explosives. Another nitrocellulose is pyroxylin, which can be made into celluloid, motion picture film, artificial leather, and lacquers for automobile finishes. Cellulose treated with acetic anhydride yields cellulose acetate, a thermoplastic polymer. This material finds applications in the manufacture of motion picture films, plastics, and cellulose acetate fiber or yarn. It is also fabricated into food wrap products and is sandwiched between sheets of glass to increase the strength of safety glass. Cellulose acetate dissolved in a volatile solvent is used in fingernail polish.

In addition to the esters of cellulose, certain ethers have become important. Methylcellulose, ethylcellulose, and carboxymethylcellulose are examples of these ethers. Methylcelluloses are used as sizing and finish for textiles, pastes, and cosmetics. Ethylcellulose has properties that make it a desirable adjunct in the manufacture of plastics, coatings, and films. It is soluble in organic solvents but very resistant to the action of alkalis. Carboxymethylcellulose is used as a protective colloid, a sizing agent for textiles, and as a builder in the manufacture of synthetic detergents.

MIXED POLYSACCHARIDES

Heparin

Heparin is a polysaccharide that possesses anticoagulant properties. It prevents the clotting of blood by inhibiting the conversion of prothrombin to thrombin. Thrombin acts as a catalyst in converting plasma fibrinogen into the fibrin clot. The structure of heparin is still uncertain but it contains a repeating unit of α-1,3 linked glucuronic acid and glucosamine, with sulfate groups on some of the hydroxyl and amino groups.

D-Glucuronic acid-2-sulfate   D-Glucosamine sulfate

α-1,3 linked repeating unit of heparin

Hyaluronic Acid

A structural polysaccharide found in higher animals is the mucopolysaccharide hyaluronic acid. It is an essential component of the ground substance, or intercellular cement, of connective tissue. Hyaluronic acid has a high viscosity and a molecular weight in the millions. The molecule consists of repeating units of D-glucuronic acid and N-acetyl-D-glucosamine joined in a β-1,3 linkage.

Chondroitin Sulfate

Chondroitin sulfates are structural polysaccharides found in the ground substance and cartilage of mammals. Its structure is similar to hyaluronic acid except that galacto-samine sulfate replaces the acetyl glucosamine.

IMPORTANT TERMS AND CONCEPTS

acetal
aldose
asymmetric carbon atom
cellulose
disaccharides
enantiomers
fructofuranose
glucopyranose
glycogen
hemiacetal

hexose
ketose
monosaccharides
mutarotation
optical activity
pentose
polarized light
polysaccharides
reducing sugar
starch

QUESTIONS

1. Define optical activity and asymmetric carbon atom.

2. Explain the difference between the *meso* form of an optically active compound and a racemic mixture.

3. Write the formula for the aldose, glyceraldehyde. Would this compound exhibit optical activity in solution? Explain.

4. Write the formula for lactic acid. Would this compound be classified as a carbohydrate? Explain.

5. What is (1) an aldose, (2) a hexose, (3) a pentose, (4) a ketose, and (5) a disaccharide?

6. Explain fully what is meant by (1) D (+) glucose, and (2) L (−) fructose.

7. How does the phenomenon of mutarotation complicate the representation of the formulas for carbohydrates? Explain with an example.

8. Explain the relation between hemiacetal ring formation and Haworth's representation of glucopyranose.

9. Illustrate the chair form of α-D-glucopyranose. What advantage does this representation have over the Haworth formula?

10. How are the α- and β- isomers of the pyranose and furanose ring forms of the sugars indicated in the structures?

11. Write an equation to illustrate the formation of an acetal or a glucoside between β-D-galacto-pyranose and the OH on carbon-4 of α-D-glucopyranose.

12. When a reducing sugar reacts with Benedict's solution, what other products are formed besides Cu_2O? Why is the formation of Cu_2O important in the test?

13. Explain the significance of the Lobry de Bruyn-von Eckenstein transformation.

14. Write the formula for sucrose and use it to explain why sucrose will not reduce Benedict's solution.

15. Write a partial polysaccharide structure that illustrates both α-1,4 and α-1,6 linkages between the monosaccharides.

16. Explain why polysaccharides such as starch, glycogen, and cellulose are not considered as reducing sugars.

17. Explain what happens when starch is hydrolyzed and how the process can be monitored.

18. How would you account for the great difference in properties between starch and cellulose?

19. Briefly discuss the versatility of cellulose.

Chapter 6

LIPIDS

The *objectives* of this chapter are to enable the student to:

1. Name and write the structures for the three most commonly occurring fatty acids in edible fats.
2. Illustrate the formation of a triglyceride from glycerol and three molecules of a fatty acid.
3. Write an equation illustrating the process of saponification.
4. Describe the separation of lipids by thin-layer and gas-liquid chromatography.
5. Distinguish between phospholipids and glycolipids.
6. Recognize the sterol nucleus in the steroid hormones.
7. Distinguish between the structures of estrone, testosterone, and cortisone.

In the living cell the lipids are involved in at least three major physiological roles. They are essential structural components of the cell membrane, are stored in the fat depots, and serve as major energy sources for man and animals. The fatty acids of lipids are oxidized in the mitochondria to form acetyl coenzyme A, which is a precursor of ATP molecules formed in the Krebs cycle, to be discussed later (p. 128). Lipids compose about 5 per cent of the organic matter of the cell, and exist in about 40 to 50 different kinds of molecules in the cell. The cells of the brain and nervous tissue are especially rich in lipids.

Lipids are characterized by the presence of fatty acids or their derivatives and by their solubility in fat solvents such as acetone, alcohol, ether, and chloroform. Chemically, lipids are composed of five main elements: carbon, hydrogen, oxygen, and occasional nitrogen and phosphorus. At present there is no generally accepted method of classification of lipids. Some schemes divide them into simple lipids, compound lipids, and steroids, but a more practical classification may be that followed in this chapter:

Fats—esters of fatty acids with glycerol.
Phospholipids—compounds that contain phosphorus, fatty acids, glycerol, and a nitrogenous compound.
Sphingolipids—compounds that contain a fatty acid, phosphoric acid, choline, and an amino alcohol, sphingosine.
Glycolipids—composed of a carbohydrate, a fatty acid, and an amino alcohol.
Steroids—high molecular weight cyclic alcohols.
Waxes—esters of fatty acids with alcohols other than glycerol.

FATTY ACIDS

Since all fats are esters of fatty acids and glycerol, it may be well to consider the composition and properties of these substances before discussing lipids in general. Fatty acids, although not lipids themselves, are sometimes classified as derived lipids, since they are constituents of all the above types of lipids. The fatty acids that occur in nature almost always have an even number of carbon atoms in their molecules. They are usually straight-chain organic acids that may be saturated or unsaturated. Some of the important fatty acids that occur in natural fats are listed in Table 6–1.

In the series of saturated fatty acids, those up to and including capric acid are liquid at room temperature. The most important saturated fatty acids are **palmitic** and **stearic acids.** They are components of the majority of the common animal and vegetable fats.

Unsaturated fatty acids are characteristic constituents of oils. **Oleic acid,** which contains one double bond, is the most common unsaturated fatty acid. Its formula is written:

$$CH_3(CH_2)_7CH{=}CH(CH_2)_7COOH$$

Ricinoleic acid is an unsaturated fatty acid characterized by the presence of a hydroxyl group and is found in castor oil. Its formula is as follows:

$$CH_3(CH_2)_5CHOHCH_2CH{=}CH(CH_2)_7COOH$$

The structure of fatty acids suggested by x-ray analysis is that of a zigzag configuration with the carbon-carbon bond forming a 109° angle.

Stearic acid

TABLE 6-1 Some Important Fatty Acids Occurring in Natural Fats

Name	Formula	Carbon Atoms	Position of Double Bonds	Occurrence
Saturated				
Butyric	C_3H_7COOH	4		Butter fat
Caproic	$C_5H_{11}COOH$	6		Butter fat
Caprylic	$C_7H_{15}COOH$	8		Coconut oil
Capric	$C_9H_{19}COOH$	10		Palm kernel oil
Lauric	$C_{11}H_{23}COOH$	12		Coconut oil
Myristic	$C_{13}H_{27}COOH$	14		Nutmeg oil
Palmitic	$C_{15}H_{31}COOH$	16		Animal and vegetable fats
Stearic	$C_{17}H_{35}COOH$	18		Animal and vegetable fats
Arachidic	$C_{19}H_{39}COOH$	20		Peanut oil
Unsaturated				
Palmitoleic (1 =)°	$C_{15}H_{29}COOH$	16	Δ9†	Butter fat
Oleic (1 =)	$C_{17}H_{33}COOH$	18	Δ9	Olive oil
Linoleic (2 =)	$C_{17}H_{31}COOH$	18	Δ9, 12	Linseed oil
Linolenic (3 =)	$C_{17}H_{29}COOH$	18	Δ9, 12, 15	Linseed oil
Arachidonic (4 =)	$C_{19}H_{31}COOH$	20	Δ5, 8, 11, 14	Lecithin

° Number of double bonds.

† Δ9 indicates a double bond between carbon 9 and 10, Δ12 between carbon 12 and 13, and so forth.

Unsaturated fatty acids exhibit a type of geometrical isomerism known as **cis-trans isomerism.** The *cis* configuration is found in nature, and the *cis*-form of unsaturated fatty acids is less stable than the *trans*-form. Oleic acid exists in the *cis* configuration; the *trans* configuration of the fatty acid is called elaidic acid.

Oleic acid (*cis*-form)

Elaidic acid (*trans*-form)

The **prostaglandins** are cyclic fatty acids first described in the 1930s by von Euler, who found them in the seminal plasma, the prostate gland, and the seminal vesicles. At first prostaglandin was thought to be a single substance secreted by the male genital tract, but more recent research has shown that there are six major kinds of prostaglandins, which function as regulators of metabolism in a number of tissues (p. 137). They are synthesized in the tissues from 20-carbon polyunsaturated fatty acids such as arachidonic acid. Since arachidonic acid is formed from linoleic acid, an essential fatty acid in the diet, this may be a major reason that polyunsaturated fatty acids like linoleic and linolenic are essential to an adequate diet. A typical example of a prostaglandin formed from arachidonic acid is E_2. (E refers to the nature of the constituents on the ring, and E_2 indicates that the compound contains two double bonds.)

E_2 (Prostaglandin)

These compounds are involved in the control of lipid metabolism (Chapter 11) and are thought to depress the action of cyclic-3′,-5′-AMP (p. 125). Their control of cyclic AMP levels may play an important role in the synthesis of protein, RNA, and even DNA.

From a nutritional standpoint, the three most commonly occurring fatty acids in edible animal and vegetable fats are palmitic, stearic, and oleic acids.

FATS

From a chemical standpoint fats are esters of fatty acids and glycerol. This combination of 3 molecules of fatty acid with 1 molecule of glycerol may be illustrated as shown in the reaction on the following page.

Tristearin is called a **simple glyceride** because all the fatty acids in the fat molecule are the same. Other examples of simple glycerides would be tripalmitin and triolein. In most naturally occurring fats, different fatty acids are found in the same molecule. These are called **mixed glycerides** and may contain both saturated and unsaturated fatty acids. The glycerides are classed as **neutral lipids** since their molecules are not charged.

Both fats and oils are esters of fatty acids and glycerol. In general, fats are solid at room temperature and are characterized by a relatively high content of saturated fatty acids. Oils are liquids that contain a high concentration of unsaturated fatty acids. A fat that contains short-chain saturated fatty acids may also exist as a liquid at room temperature.

Most of the common animal fats are glycerides that contain saturated and unsaturated fatty acids. Since the saturated fatty acids predominate, these fats are solid at room

| Glycerol | 3 molecules of stearic acid | Tristearin, a fat |

temperature. Beef fat, mutton fat, lard, and butter are important examples of animal fats. Butter fat is readily distinguished from other animal fats because of its relatively high content of short-chain fatty acids.

Glycerides that are found in vegetables usually exist as oils rather than fats. Vegetable oils such as olive oil, corn oil, cottonseed oil, and linseed oil are characterized by their high content of oleic, linoleic, and linolenic acids. Coconut oil, like butter fat, contains a relatively large percentage of short-chain fatty acids.

Reactions of Fats

Glycerol Portion. When glycerol or a liquid containing glycerol is heated with a dehydrating agent, **acrolein** is formed. Acrolein has a very pungent odor and is sometimes formed by the decomposition of glycerol in the fat of frying meats.

| Glycerol | | Acrolein |

The formation of acrolein is often used as a test for fats, since all fats yield glycerol when they are heated.

Rancidity. Many fats develop an unpleasant odor and taste when they are allowed to stand in contact with air at room temperature. The two common types of rancidity are **hydrolytic** and **oxidative**. Hydrolytic changes in fats are the result of the action of enzymes or microorganisms producing free fatty acids. If these acids are of the short-chain variety, as is butyric acid, the fats develop a rancid odor and taste. This type of rancidity is common in butter.

The most common type of rancidity is the oxidative type. The unsaturated fatty acids in fats undergo oxidation at the double bonds. The combination with oxygen results in the formation of peroxides, volatile aldehydes, ketones, and acids.

Heat, light, moisture, and air are factors that accelerate oxidative rancidity. The prevention of rancidity of lard and vegetable shortenings that are used in the manufacture of crackers, pretzels, pastries, and similar food products has long been an important problem. Modern packaging has helped considerably in this connection, although a more important contribution has been the development of "antioxidants." These compounds usually contain phenolic groups in their structure, i.e., tocopherol, or vitamin E (p. 101), which effectively inhibits the autoxidation of unsaturated fatty acids. The majority of the vegetable shortenings on the market as well as certain brands of lard are protected from rancidity by the addition of antioxidants.

Hydrogenation. It has already been stated that the main difference between oils and fats is the number of unsaturated fatty acids in the molecule. Vegetable oils may be converted into solid fats by the addition of hydrogen to the double bonds of the unsaturated fatty acids.

Hydrolysis. Fats may be hydrolyzed to form free fatty acids and glycerol by the action of acid, alkali, superheated steam, or the enzyme lipase. In hydrolysis of a fat, the 3 water molecules (that were split out when the 3 fatty acid molecules combined with 1 glycerol molecule in an ester linkage to make the fat molecule) are restored with the resultant splitting of the fat into glycerol and fatty acids. Commercially, fats are a cheap source of glycerol for use in the manufacture of high explosives and pharmaceuticals. For this purpose the fat is hydrolyzed with superheated steam and a reagent which contains naphthalene, oleic acid, and sulfuric acid. The advantage of this method is that glycerol is readily separated from the fatty acids.

Saponification

Hydrolysis by an alkali is called **saponification,** and produces glycerol and salts of the fatty acids that are called soaps. In the laboratory, fats are usually saponified by an alcoholic solution of an alkali. The fats are more soluble in hot alcohol and the reaction is therefore more rapid. **Soaps** may be defined as metallic salts of fatty acids. The saponification of a fat may be represented as follows:

$$CH_2{-}O{-}\overset{\displaystyle O}{\overset{\|}{C}}{-}C_{17}H_{35}$$
$$|$$
$$CH{-}O{-}\overset{\displaystyle O}{\overset{\|}{C}}{-}C_{17}H_{35} + 3NaOH \rightarrow CHOH + 3C_{17}H_{35}\overset{\displaystyle O}{\overset{\diagup}{C}}{-}ONa$$
$$|$$
$$CH_2{-}O{-}\overset{\displaystyle O}{\overset{\|}{C}}{-}C_{17}H_{35}$$

CH₂OH | CHOH | CH₂OH — Glycerol; Sodium stearate (soap); Tristearin

Sodium salts of fatty acids are known as **hard soaps,** whereas potassium salts form **soft soaps.** The ordinary cake soaps used in the home are sodium soaps. Yellow laundry soap contains resin, which increases the solubility of soap and its lathering properties and has some detergent action. White laundry soap in the form of bars, soap chips, or powdered soap contains sodium silicate and a water-softening agent such as sodium carbonate or sodium phosphate. Tincture of green soap, commonly used in hospitals, is a solution of potassium soap in alcohol. When sodium soaps are added to hard water, the calcium and magnesium ions present replace sodium to form insoluble calcium and magnesium soaps. The familiar soap curd formed in hard water is due to these **insoluble soaps.**

Detergents. These compounds are a mixture of the sodium salts of the sulfuric acid esters of lauryl and cetyl alcohols. They may be used in hard water because they do not form insoluble compounds with calcium and magnesium.

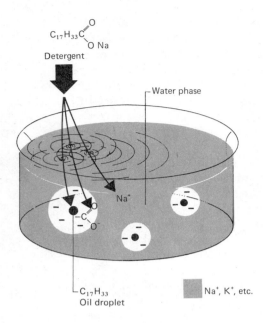

Figure 6-1 A diagram illustrating the action of a soap or detergent in stabilizing an oil and water emulsion.

Extensive research on new detergents and emulsifying agents has resulted in the development of several hundred products possessing a variety of properties, including biodegradability. At present over four billion pounds of detergents are sold each year with synthetic detergents, called **syndets,** outselling soap by more than four to one.

Soaps and detergents are emulsifying agents that can convert a mixture of oil and water into a permanent emulsion. The **cleansing power** of soaps and detergents is related to their action as emulsifying agents and to their ability to lower surface tension. By emulsifying the grease or oily material that holds the dirt on the skin or clothing, one can rinse off the particles of grease and dirt with water. The ability of soaps and detergents to break or stabilize oil and water emulsion has been given the name **"detergency."** This property may be illustrated with a diagram (Fig. 6–1).

The hydrocarbon portion of the soap molecule tends to dissolve in the oil droplets, while the carboxyl group is strongly attracted to the aqueous phase. As a result of this phenomenon, each oil droplet is negatively charged and tends to repel other oil droplets, resulting in a stable emulsion. **Detergents** are composed of a hydrophilic group similar to the carboxyl group and a hydrocarbon chain.

Analysis of Lipids

For many years the chemical analysis of lipid mixtures has been most difficult. Determination of the **saponification number** yielded a rough measure of molecular weight, the **iodine number** was used for the content of unsaturated fatty acids, the **Reichert-Meissl number** for the content of volatile fatty acids, and the **acetyl number** for the amount of hydroxy fatty acids. With the exception of the iodine number that is sometimes still used, these determinations have been replaced by more sensitive methods. Thin-layer chromatography and gas-liquid chromatography are presently the methods of choice for the analysis of lipids.

Thin-layer chromatography is carried out on a thin, uniform layer of silica gel spread on a glass plate and activated by heating in an oven (100° to 250°C). Samples of lipid material in the proper solvent are spotted along one edge of the plate with micropipettes.

After evaporation of the solvent, the plates are placed vertically in a covered glass tank which contains a layer of suitable solvent on the bottom. Within a few minutes the lipids are separated by the solvent rising through the thin layer carrying the spots to different locations on the silica gel by a combination of adsorption on the gel and varying distribution in the solvent system. The plates are removed, dried, and sprayed with various detection agents to visualize the lipid components. An example of the thin-layer chromatographic separation of several of the phospholipids is shown in Figure 6–2. This technique is very sensitive and can be made quantitative by removing the spots and measuring the concentration of the component by gas-liquid chromatography.

Gas-liquid chromatography is another powerful tool of the lipid chemist. Any substance that is volatile or can be made into a volatile derivative, for example fatty acids being converted into their methyl ester, can be separated and analyzed by this technique. The volatile substance is injected into a long column which contains a nonvolatile liquid on a finely divided inert solid. The column is heated and the volatile material is carried through the tube by an inert gas such as helium. Separation depends on the difference in vapor pressure and the partition coefficients of the components in the nonvolatile liquid. As the fractionated components reach the end of the column, they pass over a detection device that is extremely sensitive to differences in organic material carried by the gas, and it records the changes in the gas flow as peaks on a recorder chart. By the use of helium gas alone as the control, and known lipid components as standards to determine the position and area under the peaks, quantitative analysis of lipids can be achieved.

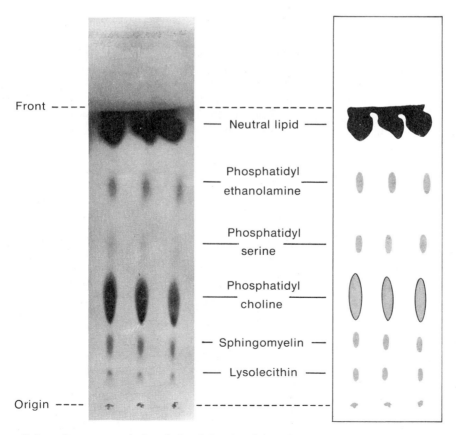

Front

— Neutral lipid —

Phosphatidyl
ethanolamine

Phosphatidyl
serine

Phosphatidyl
choline

— Sphingomyelin —

— Lysolecithin —

Origin

FIGURE 6-2   Separation of phospholipids by thin layer chromatography.

PHOSPHOLIPIDS

The phospholipids are found in all animal and vegetable cells. They are composed of glycerol, fatty acids, phosphoric acid, and a nitrogen-containing compound. More specifically, they are esters of **phosphatidic acid** with choline, ethanolamine, serine, or inositol (hexahydroxycyclohexane).

Phosphatidyl Choline, or Lecithins

The lecithins are esters of phosphatidic acid and choline.

L-α-Phosphatidic acid   α-Lecithin

The formula is written with the fatty acid on the left side of the central, or β-carbon, to indicate optical activity and an asymmetric carbon atom. Naturally occurring phosphatides have the L form and may contain at least five different fatty acids; however, the β-carbon usually is attached to an unsaturated fatty acid. In addition, the formula indicates that lecithin exists in the dissociated state since phosphoric acid is a fairly strong acid and choline is a strong base. **Choline** is a quaternary ammonium compound whose basicity in aqueous solution is similar to that of KOH.

The lecithins are constituents of brain, nervous tissue, and egg yolk. From a physiological standpoint they are important in the transportation of fats from one tissue to another and are essential components of the protoplasm of all body cells. In industry lecithin is obtained from soybeans and finds wide application as an emulsifying agent.

If the oleic acid on the central carbon atom of lecithin is removed by hydrolysis, the resulting compound is called **lysolecithin.** Disintegration of the red blood cells, or hemolysis, is caused by intravenous injection of lysolecithin. The venom of snakes such as the cobra contains an enzyme capable of converting lecithins into lysolecithins, which accounts for the fatal effects of the bite of these snakes. A few insects and spiders produce toxic effects by the same mechanism.

Phosphatidyl Ethanolamine, or Cephalins

The cephalins are found in brain tissue and are essentially mixtures of phosphatidyl ethanolamine and phosphatidyl serine.

Basic structure

For Phosphatidyl serine,

R^1 is —CH_2CHNH_2
                    |
                  COOH

For Phosphatidyl ethanolamine,

R^1 is —$CH_2CH_2NH_2$

The cephalins are involved in the blood-clotting process and are therefore essential constituents of the body.

Other types of phospholipids related to the lecithins and cephalins are the phosphatidyl inositols and the plasmologens, which are derivatives of phosphatidyl ethanolamine. These compounds are not completely characterized at present but are found in brain, heart, and liver tissue.

SPHINGOLIPIDS

Sphingomyelins

The sphingomyelins differ chemically from the lecithins or cephalins. The common structure in both sphingolipids and glycolipids is **ceramide,** which is composed of sphingosine and a fatty acid (R).

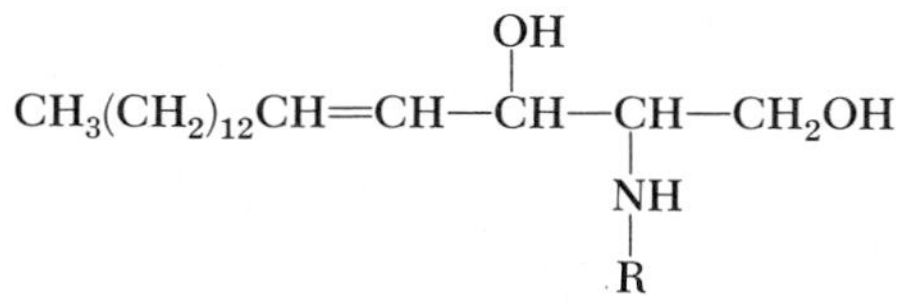

$$CH_3(CH_2)_{12}CH{=}CH{-}\underset{\underset{R}{\overset{|}{NH}}}{\overset{\overset{OH}{|}}{CH}}{-}CH{-}CH_2OH \qquad Ceramide{-}O{-}\underset{\underset{O^-}{\overset{||}{P}}}{\overset{\overset{O}{||}}{P}}{-}O{-}CH_2CH_2\overset{+}{N}(CH_3)_3$$

Ceramide 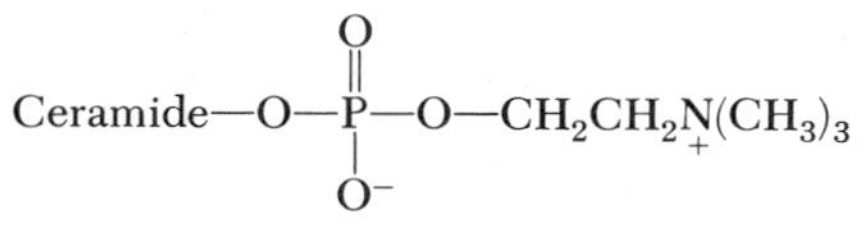 Sphingomyelin

The sphingomyelins are found in large amounts in brain and nervous tissue and are essential constituents of the protoplasm of cells.

GLYCOLIPIDS

The glycolipids are compound lipids that contain a carbohydrate.

Cerebrosides

These lipids are often called **cerebrosides** because they are found in brain and nervous tissue. They are composed of a carbohydrate and a ceramide. There are four different cerebrosides, each containing a different fatty acid. **Kerasin** contains lignoceric acid ($C_{23}H_{47}COOH$), **phrenosin** a hydroxy lignoceric acid, **nervon** an unsaturated lignoceric acid with one double bond called nervonic acid, and **oxynervon,** a hydroxy derivative of nervonic acid. The carbohydrate in these lipids is usually galactose, although glucose is sometimes present. The structure of a typical cerebroside may be represented as follows:

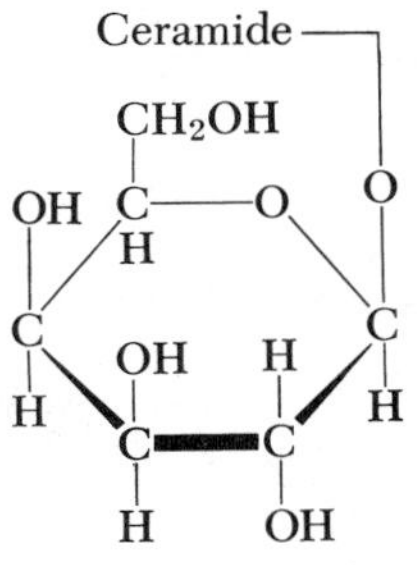

Cerebroside

STEROIDS

The steroids are derivatives of cyclic alcohols of high molecular weight that occur in all living cells. The lipid material from tissue that is not saponifiable by alkaline hydrolysis contains the steroids. The parent hydrocarbon compound for all the steroids is the cyclopentanophenanthrene nucleus, also called the sterol nucleus. This structure is an integral part of the cholesterol molecule which may be used to illustrate the lettering system for the rings and the number system for the carbon atoms.

Sterol nucleus   Cholesterol structure designation

The most common sterol is **cholesterol,** which is found in brain and nervous tissue and in gallstones. The structure of cholesterol is shown as follows (each ring is completely saturated and where there is a double bond in the ring it is specifically designated):

Cholesterol

Cholesterol reacts with acetic anhydride and sulfuric acid in a dry chloroform solution to yield a green color. This is called the **Liebermann-Burchard reaction** and is the basis for both qualitative detection and quantitative methods for cholesterol.

TOPIC OF CURRENT INTEREST

THE OFFSPRING OF CHOLESTEROL

Cholesterol is a much-publicized molecule at the present time. For the past few years we have been told repeatedly to eat less of cholesterol-rich foods such as eggs, in order to help prevent atherosclerosis and hardening of the arteries. Replacing saturated fats in the diet with those containing unsaturated fatty acids is supposed to reduce the cholesterol concentration of the blood. More recently, our intake of neutral fats has been implicated in the increase of cholesterol in the body. Since cholesterol can be synthesized in the liver and other tissues from simple two-carbon compounds such as acetyl coenzyme A, it must have some function in the body, and must be maintained at a physiological level in the blood for some reason.

The reasons are evident when we examine the many and varied offspring of cholesterol. Biochemically, cholesterol is a sterol and is classed as a steroid. It is closely related to other animal sterols such as 7-dehydrocholesterol (a precursor of vitamin D), dihydrocholesterol (present with cholesterol in tissues), and coprosterol, which is present in feces. Two important plant sterols are sitosterol and stigmasterol.

Cholesterol is the precursor of many other steroids in animal tissues, including bile acids, male sex hormones, female sex hormones, adrenal corticosteroids, vitamin D, saponins, and the cardiac glycosides. The majority of these steroids will be discussed in greater detail in other sections of the text, but brief mention should be made of their importance to the body. The bile salts and bile acids are detergent-like compounds which aid in emulsification and absorption of lipids in the intestine and which activate the lipase involved in the digestion of fats. The male sex hormones, or androgens, are formed in the testes, and are responsible for the development of the secondary sexual characteristics of the male, beginning in puberty. The female sex hormones, or estrogens (including progesterone), are essential for the normal menstrual cycle, the development of secondary sexual characteristics in the adult female, and the alteration of the menstrual cycle during pregnancy. The adrenal corticosteroids consist of several important hormones such as cortisone, cortisol, aldosterone, and derivatives of corticosterone. Compounds having vitamin D activity are vitamins D_2 and D_3, which are related to ergosterol and 7-dehydrocholesterol. Saponins are represented by digitonin, a plant glycoside of a steroid in which the sugar residues are attached to the OH group at carbon 3. Digitonin forms insoluble addition compounds with cholesterol and is used in methods for quantitation of the sterol. An important cardiac glycoside is digitoxigenin, which is related to two important drugs, digoxin and digitoxin, which are used in the treatment of cardiac disease.

Perhaps even this brief account of steroids dependent on cholesterol as a precursor will focus our attention on the benefits of the sterol. Although abnormally large concentrations in the blood are related to the condition of atherosclerosis, the presence of the cholesterol molecule is vital to a wide range of body functions.

BILE SALTS

The bile salts are natural emulsifying agents found in the bile, a digestive fluid formed by the liver. Cholesterol and bile pigments are also important constituents of the bile. Bile is stored in the gall bladder and released at intervals to assist in the digestion and absorption of fats. **Cholic acid** and **deoxycholic acid** are the major bile acids that are combined with glycine or taurine by an amide linkage to form bile salts such as glycocholate or taurocholate.

Cholic acid: R^1 is OH and R^2 is OH
Deoxycholic acid: R^1 is OH and R^2 is H
Glycocholic acid: R^1 is NH_2CH_2COOH (Glycine)
Taurocholic acid: R^1 is $NH_2CH_2CH_2SO_3H$ (Taurine)

Basic structure

Chenodeoxycholic acid with hydroxyl groups at positions 3 and 7 in rings A and B, and **lithocholic acid** with a single hydroxyl group at position 3, are also found in human bile.

HORMONES OF THE ADRENAL CORTEX

The cortex of the adrenal gland is an endocrine gland (Fig. 6–3) which produces a group of hormones with important physiological functions. If the gland exhibits decreased function, as in Addison's disease, electrolyte and water balance are abnormal, carbohydrate and protein metabolisms are adversely affected, and the patient is more

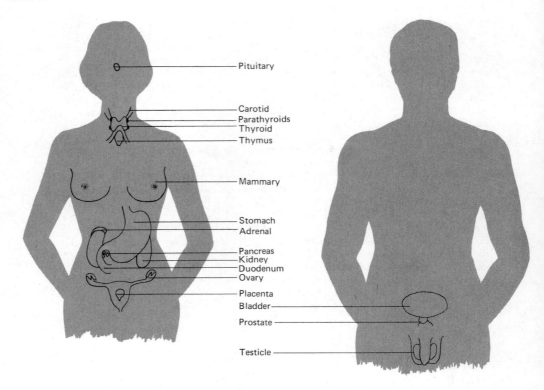

FIGURE 6-3   The location of the endocrine glands in the body.

sensitive to cold and stress. Typical steroid hormones of the gland are represented as follows:

Corticosterone was the original name of the first adrenal cortical hormone, which accounts for the naming of other hormones as derivatives of this compound. Three major types of adrenal cortical hormones illustrate the relation of structure to physiological activity.

1. Compounds containing an oxygen on the C-11 position (C—OH, or C═O) exhibit greatest activity in carbohydrate and protein metabolism. Examples are **corticosterone, cortisone,** and **cortisol.**

2. Hormones without an oxygen on the C-11 position have their greatest effect on electrolyte and water metabolism. Examples are **11-deoxycorticosterone** and **11-deoxycortisol.**

3. **Aldosterone** is the only compound without a methyl group at C-18. It is replaced by an aldehyde group that can exist in the aldehyde form or in the hemiacetal form. Aldosterone has a very potent effect on electrolytes and is called a **mineralocorticoid.** In higher doses it also acts on carbohydrate and protein metabolism.

Corticosterone, cortisol, and aldosterone are the major hormones found in the blood, with **cortisol** exerting the greatest effect on carbohydrate and protein metabolism, and aldosterone on the body fluid electrolytes.

When first tried clinically, **cortisone** stimulated considerable excitement in the treatment of rheumatoid arthritis. However, it was subsequently found that the original symptoms would reappear after a period of treatment and that unwanted side effects resulted from the use of this steroid. The pharmaceutical industry prepared and tried many closely related compounds to increase the potency and decrease the side effects of the drug (Chapter 15, p. 192).

FEMALE SEX HORMONES

The female sex hormones are steroid in structure and are formed in the ovaries, which are glands lying on the sides of the pelvic cavity (Fig. 6–3). Follicles and corpus lutea in different stages of development are located in the cortex of the ovary and form hormones that regulate the **estrus** or **menstrual cycle** and function in pregnancy (Fig. 6–4). Follicular hormones are also responsible for the development of the secondary sexual characteristics that occur at puberty.

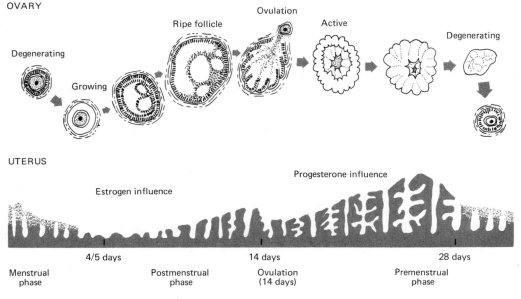

FIGURE 6–4   The sequence of events in the menstrual cycle.

Hormones of the Follicle

The liquid within the follicle contains at least two hormones, known as **estrone** and **estradiol.** Estrone (theelin) was the first hormone to be isolated from the follicular liquid, but estradiol (dihydrotheelin) is more potent than estrone and may be the principal hormone.

Estrone   Estradiol

These two compounds are excreted in the urine in increased amounts during pregnancy.

The Hormone of the Corpus Luteum

The hormone produced by the corpus luteum is called **progesterone.** In the body progesterone is converted into **pregnanediol** by reduction before it is excreted in the urine. These two compounds are similar to the estrogens in chemical structure.

Progesterone   Pregnanediol

The main function of progesterone is the preparation of the uterine endometrium for implantation of the fertilized ovum. If pregnancy occurs, this hormone is responsible for the retention of the embryo in the uterus and for the development of the mammary glands prior to lactation. In the normal menstrual cycle the administration of progesterone inhibits ovulation, a property used in the development of "the pill" (p. 192).

MALE SEX HORMONES

The male sex hormones are produced by the testes, which are two oval glands located in the scrotum of the male (Fig. 6–3). Small glands in the testes form spermatozoa, which are capable of fertilizing a mature ovum. Between the cells that manufacture spermatozoa are the **interstitial cells,** which produce a hormone called **testosterone.** This hormone is probably converted into other compounds such as **androsterone** before being excreted in the urine. **Dehydroandrosterone** has also been isolated from male urine but is much less active than the other two hormones. The male sex hormones, or **androgens,** have structures similar to the estrogens.

Testosterone   Androsterone

The main function of the androgens in man is the development of masculine sexual characteristics, such as deepening of the voice, the growth of a beard, and distribution of body hair at puberty. They also control the function of the glands of reproduction (seminal vesicles, prostate, and Cowper's gland).

WAXES

Waxes are simple lipids that are esters of fatty acids and high molecular weight alcohols. Fatty acids such as myristic, palmitic, and carnaubic are combined with alcohols that contain from 12 to 30 carbon atoms. Common, naturally occurring waxes are **beeswax, lanolin, spermaceti,** and **carnauba wax.** Beeswax is found in the structural part of the honeycomb. Lanolin, from wool, is the most important wax from a medical standpoint, since it is widely used as a base for many ointments, salves, and creams. Spermaceti, obtained from the sperm whale, is used in cosmetics, some pharmaceutical products, and in the manufacture of candles. Carnauba wax is obtained from the carnauba palm and is widely used in floor waxes and in automobile and furniture polishes.

IMPORTANT TERMS AND CONCEPTS

cephalin
cerebroside
cholesterol
detergents
estrone
lecithin

rancidity
saponification
steroids
testosterone
triglyceride
unsaturated fatty acid

QUESTIONS

1. Name and write the formulas for an unsaturated and a hydroxy-containing fatty acid.

2. Discuss the formation and functions of the prostaglandins.

3. Are most commonly occurring fats composed of simple glycerides or mixed glycerides? Explain.

4. Write the formula for a triglyceride that would exist as a solid at room temperature and for one that would exist as an oil.

5. Explain how you would test for the presence of glycerol in the laboratory and why this test can be used as a general test for fats.

6. What type of rancidity occurs in common shortenings? How is this prevented? Explain.

7. Write an equation illustrating the process of saponification. Name all compounds in the equation.

8. How do insoluble soaps, soft soaps, and hard soaps differ from each other?

9. Explain the cleansing power of soaps and detergents.

10. What analytical procedure would you employ to separate a mixture of phospholipids and neutral lipids into its components? Explain.

11. Write the formula for a typical phosphatidyl ethanolamine. What function does this compound serve in the body?

12. Discuss the importance of cholesterol in the body.

13. Explain the relation between the structure of the adrenal cortical steroids and their physiological function.

14. What is the major structural difference between aldosterone and the other steroid hormones of the adrenal cortex?

15. Write the structure for estrone and indicate how it differs from the male sex hormones.

Chapter 7

VITAMINS AND COENZYMES

The *objectives* of this chapter are to enable the student to:

1. Define and describe the properties of vitamins.
2. Explain the relationship between water-soluble vitamins and coenzymes.
3. Explain the difference between a coenzyme, a prosthetic group, and an apoenzyme.
4. Describe the chemical nature of a specific coenzyme and its function in the body.
5. Describe the relation between vitamin A and Δ^{11} *cis*-retinal.
6. Describe the formation of vitamin D_2 from ergosterol.

As we have seen, the major components of living cells are proteins, nucleic acids, carbohydrates, and lipids. In addition, cells contain organic compounds called **vitamins,** which function in trace amounts. The interest in vitamin therapy, chemistry, and function began in antiquity, when it was recognized that diet was related to disease. In the first four decades of this century, biochemical research was mainly concerned with nutrition, including the study of vitamins. In the 1930s and 40s, when vitamins became available in adequate quantities, the vitamin deficiency diseases could be treated, and their occurrence decreased markedly in many countries. Modern nutritional practices include adequate daily vitamin dosage, which has in some instances led to uncontrolled self-overdosing and a condition of **hypervitaminosis.** Excessive ingestion of vitamin A produces nausea, vomiting, skin irritations, and mental disturbances, while overdoses of vitamin D lead to increased blood calcium, depression of brain function, and kidney damage.

COENZYMES

In early studies of the enzymes of yeast it was observed that dialysis of a solution of yeast inactivated the enzymes. When the dialyzed material was added to the enzymes, they again exhibited activity. The cofactor in the dialysate was called a **coenzyme.** Since that time several coenzymes have been discovered, and they have been found to consist of small organic molecules. If the organic molecule, or nonprotein portion, is readily separated from the enzyme, it is called a coenzyme. If it is firmly attached to the protein portion of the enzyme, it is called a **prosthetic group.** Most enzymes may therefore be considered as conjugated proteins composed of an inactive protein molecule called the **apoenzyme** combined with the prosthetic group or coenzyme. The complete, conjugated, active molecule is called a **holoenzyme.**

WATER-SOLUBLE VITAMINS

Vitamins or derivatives of vitamins serve in intermediary metabolism as coenzymes or prosthetic groups in enzymatic reactions involving oxidation, reduction, and decarboxylation. The water-soluble B vitamins contain several vitamins that exhibit the properties of coenzymes.

Vitamin B_1

Vitamin B_1, or thiamine, contains a pyrimidine ring and a sulfur-containing thiazole ring.

Thiamine chloride

A deficiency of the vitamin in the diet results in a disease called **polyneuritis** in animals and **beriberi** in man. The peripheral nerves of the body are involved, with muscle cramps, numbness of the extremities, pain along the nerves, and eventually atrophy of muscles, edema, and circulatory disturbances occurring in the body.

Yeast, whole-grain cereals, eggs, and pork are good sources of the vitamin. Thiamine occurs free in cereal grains, but occurs as the coenzyme, thiamine pyrophosphate, in yeast and meat.

Thiamine pyrophosphate (cocarboxylase) (TPP)

Cocarboxylase functions in the oxidative decarboxylation of pyruvic acid to form acetyl coenzyme A and carbon dioxide (p. 129). The thiazole ring is the active site of this function, with the hydrogen atom dissociating as a proton from carbon-2 and the formation of a carbanion. The carbanion structure then reacts with pyruvic acid to form CO_2 and acetaldehyde. The essential reactions may be outlined as follows:

Thiazole ring carbanion   Intermediate   Acetaldehyde

Cocarboxylase is therefore essential in the conversion of pyruvic acid to acetaldehyde. If this reaction does not occur at a normal rate, pyruvic acid may accumulate in the blood and tissues and give rise to the neuritis that is common in thiamine deficiency. **Thiamine pyrophosphate, TPP,** also serves as a coenzyme for enzymes such as α-keto acid oxidase, phosphoketolase, and transketolase.

Riboflavin

Riboflavin, or vitamin B_2, is composed of a pentose alcohol, ribitol, and a pigment, flavin.

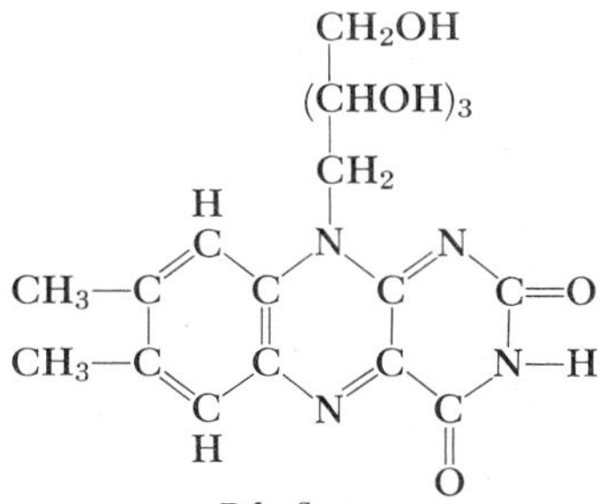

Riboflavin

A deficiency of vitamin B_2 in the diet of animals such as the rat, dog, and chicken causes lack of growth, loss of hair, and cataracts of the eyes. Lack of the vitamin in the human affects vision and causes inflammation of the cornea, and sores and cracks in the corners of the mouth.

Foods rich in riboflavin are yeast, liver, eggs, and leafy vegetables. Milled cereal products lose both their thiamine and vitamin B_2, and at present there is a trend toward the fortification of white flour with these two vitamins.

The vitamin functions as a coenzyme; in fact, it occurs in foods as a component of two flavin coenzymes, FMN and FAD. The structures of **flavin mononucleotide, FMN,** and **flavin adenine dinucleotide, FAD,** are represented as follows:

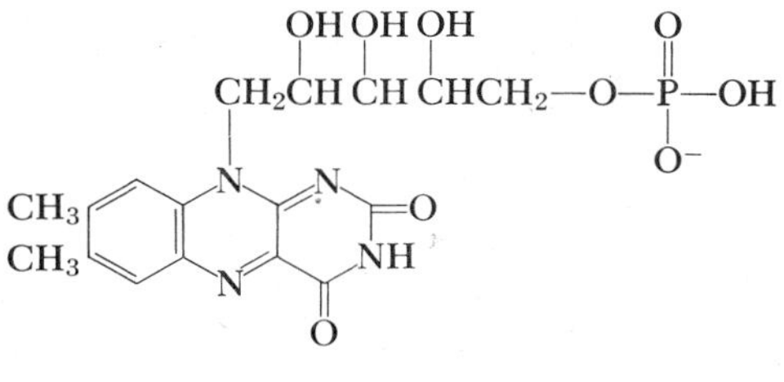

Flavin mononucleotide (FMN)

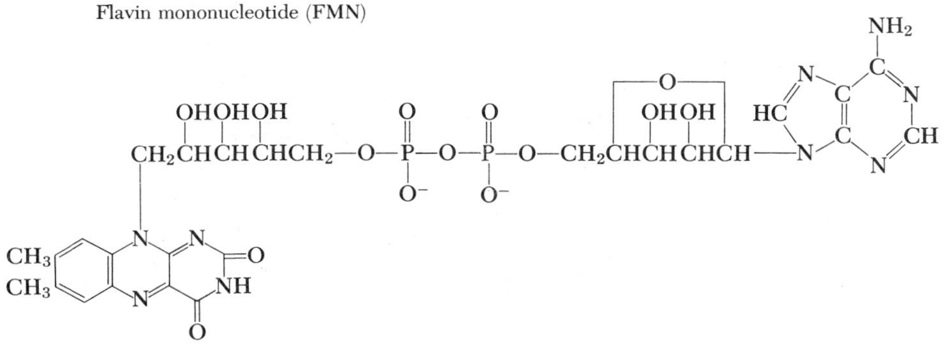

Flavin adenine dinucleotide (FAD)

Both FMN and FAD serve as coenzymes for a group of enzymes which catalyze oxidation-reduction reactions. Glutathione reductase, succinic dehydrogenase, and D-amino acid oxidase are examples of these enzymes. The flavin portion of the molecule is the active site for the oxidation-reduction reactions.

Nicotinic Acid and Nicotinamide

These two compounds have comparatively simple structures and as vitamins are called **niacin.**

Nicotinic acid   Nicotinamide

A deficiency of niacin in the diet results in **pellagra** in man and blacktongue in dogs. Pellagra is a disease characterized by skin lesions that develop on parts of the body that are exposed to sunlight. A sore and swollen tongue, loss of appetite, diarrhea, and nervous and mental disorders are typical symptoms of the disease.

Liver, lean meat, and yeast are good sources of niacin, whereas corn, molasses, and fat meat are very poor sources. Pellagra used to be more prevalent in the South, where the latter three foods are major constituents of the diet.

Niacin is an essential component of two important coenzymes, **nicotinamide-adenine dinucleotide, NAD,** and **nicotinamide-adenine dinucleotide phosphate, NADP.**

Nicotinamide-adenine
dinucleotide (NAD)

Nicotinamide-adenine
dinucleotide phosphate (NADP)

The nicotinamide portion of NAD and NADP is involved in the mechanism of the oxidation-reduction reactions with which these coenzymes are involved.

TOPIC OF CURRENT INTEREST

NAD AND NADP—IMPORTANT COENZYMES IN THE BODY AND THE LABORATORY

NAD and NADP are also referred to as **pyridine coenzymes,** since nicotinamide is a derivative of pyridine. These compounds function as the coenzymes of oxidoreductases, and are called pyridine-linked dehydrogenases. Over 200 of these dehydrogenases are known, and they function in many types of metabolic reactions. Dehydrogenase systems utilizing these coenzymes include those involved in the oxidation and reduction of lactate, malate, ethanol, glycerol-3-phosphate, isocitrate, and glucose-6-phosphate. They catalyze the general reactions:

$$\text{Reduced substrate} + \begin{matrix} \text{NAD}^+ \\ \text{or} \\ \text{NADP}^+ \end{matrix} \rightleftharpoons \text{Oxidized substrate} + \begin{matrix} \text{NADH} \\ \text{or} \\ \text{NADPH} \end{matrix} + \text{H}^+$$

The reactions involve the transfer of two hydrogen atoms from the substrate, one in the form of a hydride ion (H^-) to the 4 position of the oxidized nicotinamide ring; the other hydrogen atom is removed from the substrate as a free H^+ ion. This may be illustrated in the lactate dehydrogenase reaction as follows:

Lactic acid   NAD or NADP   Pyruvic acid   NADH or NADPH

These coenzymes are bound relatively loosely to the dehydrogenase protein molecule during the catalytic cycle; they serve more as substrates than as prosthetic groups. In essence, they act as electron acceptors during the enzymatic removal of hydrogen atoms from substrate molecules such as lactate.

In the cell, the pyridine-linked dehydrogenases are localized according to their function in metabolic reactions. Lactate dehydrogenase, for example, is found in the cytoplasm, β-hydroxybutyrate dehydrogenase in the mitochondria, and malate dehydrogenase in both compartments. NAD is present in greater concentration than NADP in animal cells. In the liver about 60 per cent of the NAD is located in the mitochondria, versus 40 per cent in the cytoplasm.

Since the dehydrogenases are involved in a multitude of metabolic reactions, methods for their assay in the biochemistry and clinical laboratories are of interest to investigators. Three characteristic changes that are useful for measuring the activity of these dehydrogenases result from the enzymatic reduction of NAD^+ and $NADP^+$. Both coenzymes absorb strongly in the ultraviolet at 260 nm, whereas when they are reduced a new absorption peak occurs at 340 nm. This appearance (or disappearance) of the 340 nm absorbance may be used to follow the reactions catalyzed by these coenzymes. Since an H^+ is released into the media when a substrate is reduced, a change of pH may serve as a measure of enzyme activity. In addition, a change in fluorescence occurs when the coenzymes are reduced and may be used in the dehydrogenase assay. The measurement that is most commonly used in the laboratory is the change in absorbance at 340 nm.

$$NAD^+ \quad \quad NADH$$
$$or \quad \rightleftharpoons \quad or$$
$$NADP^+ \quad \quad NADPH$$

Zero absorption of   Strongly absorbs
light at 340 nm   light at 340 nm

Clinically important enzymes are routinely assayed by a direct reaction involving NAD or NADP as coenzymes or as a series of reactions that end with such a direct reaction. By the proper coupling of enzymatic reactions, clinical constituents such as blood glucose can be quantitatively measured by reading the change of absorbance at 340 nm. These coenzyme reactions are used so frequently in the laboratory that spectrophotometers set to read at a single wavelength (340 nm) are available commercially.

Pyridoxine

The original name for this vitamin was **vitamin B$_6$**, which is a general name for **pyridoxine** and two closely related compounds, **pyridoxal** and **pyridoxamine.** These compounds, like nicotinic acid, are pyridine derivatives.

Pyridoxine   Pyridoxal   Pyridoxamine

A deficiency of vitamin B$_6$ in the diet of young rats results in a dermatitis called **acrodynia,** which is characterized by swelling and edema of the ears, nose, and paws. Pigs, cows, dogs, and monkeys exhibit central nervous system disturbances on a pyridoxine-deficient diet.

Vitamin B$_6$ is widely distributed in nature, with yeast, eggs, liver, cereals, legumes, and milk serving as good sources. The phosphate derivatives of pyridoxal and pyridoxamine occur in vitamin B$_6$ sources and serve as the coenzyme forms of the vitamin.

Pyridoxal phosphate   Pyridoxamine phosphate

Pyridoxal phosphate is the major coenzyme for several enzymes involved in amino acid metabolism. Processes such as transamination, decarboxylation, and racemization of amino acids require pyridoxal phosphate as a cofactor. The active site in pyridoxal phosphate is the aldehyde group and the adjacent hydroxyl group. The functional mechanism of pyridoxal and pyridoxamine phosphates in transamination is described in Chapter 12.

Pantothenic Acid

Pantothenic acid is an amide of dihydroxydimethylbutyric acid and β-alanine, an unnatural form of the amino acid.

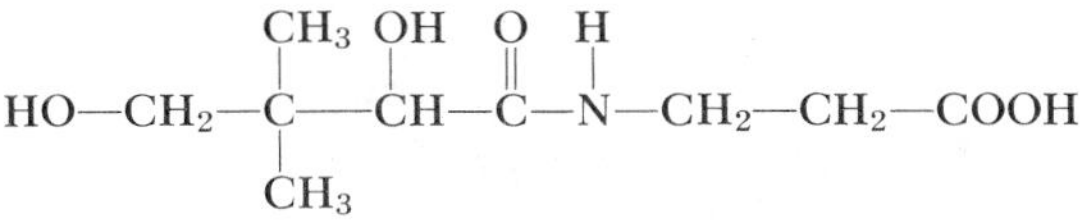

$$\text{HO—CH}_2\text{—}\underset{\underset{\text{CH}_3}{|}}{\overset{\overset{\text{CH}_3}{|}}{\text{C}}}\text{——CH}\underset{|}{\overset{|}{\text{—}}}\overset{\text{OH}}{}\overset{\overset{\text{O}}{\|}}{\text{C}}\text{—}\overset{\text{H}}{\underset{|}{\text{N}}}\text{—CH}_2\text{—CH}_2\text{—COOH}$$

Pantothenic acid

Many animals show deficiency symptoms on diets lacking pantothenic acid; for example, the rat fails to grow, and exhibits a dermatitis, graying of hair, and adrenal cortical failure. In recent dietary research on pantothenic acid deficiency in man, such symptoms as emotional instability, gastrointestinal tract discomfort, and a burning sensation in the hands and feet have been observed.

Pantothenic acid is so widespread in nature that it was named from the Greek word *pantos,* meaning everywhere. Yeast, eggs, liver, kidney, and milk are good sources. The coenzyme form of this vitamin is known as **coenzyme A.**

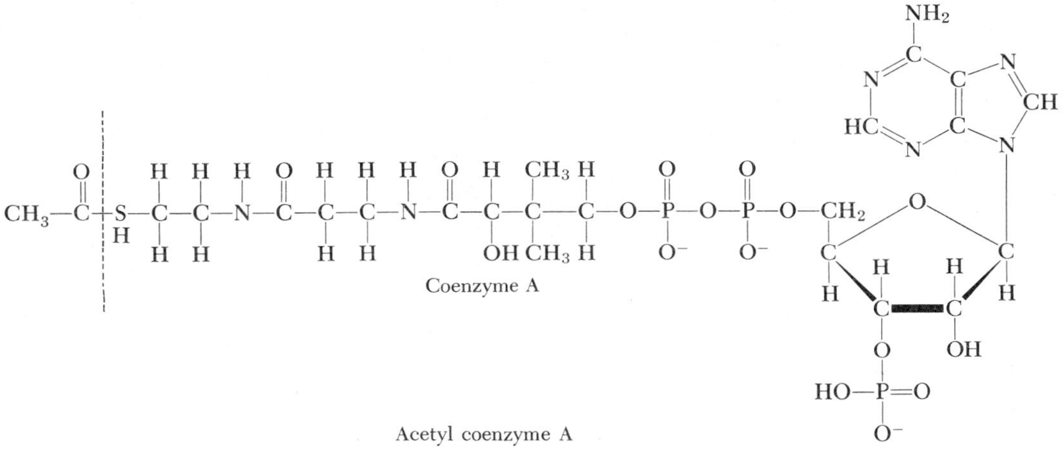

Coenzyme A

Acetyl coenzyme A

The functional group of the coenzyme is the —SH group, resulting in the abbreviated form CoASH. In biological systems it functions mainly as **acetyl CoA,** and is involved in acetylation reactions, synthesis of fats, synthesis of steroids, and the metabolic reactions that will be discussed in subsequent chapters. The formation of acetyl CoA involves a reaction of the functional —SH group with acetic acid or an acetate group. Acetyl CoA is also important as a source of acetate for the Krebs cycle.

Folic Acid

Folic acid is a complex molecule consisting of three major parts: a yellow pigment called a pteridine, *p*-aminobenzoic acid, and glutamic acid. Its composition led to the name **pteroylglutamic acid** and its structure may be represented as follows:

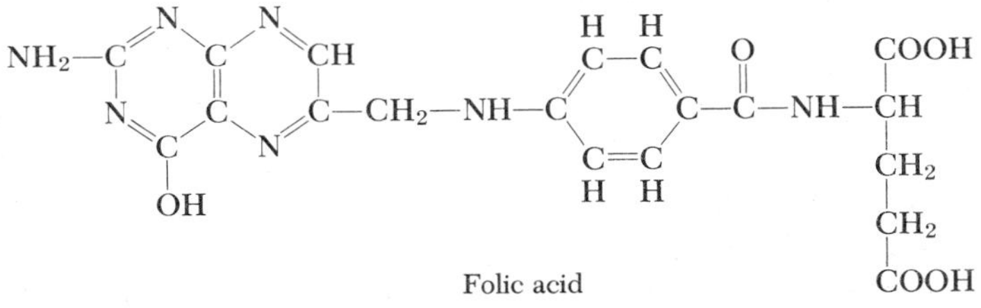

Folic acid

A lack of this vitamin in the diets of young chickens and monkeys causes anemia and other blood disorders. Recently favorable clinical results have been reported in man following the use of folic acid in **macrocytic anemias,** which are characterized by the presence of giant red corpuscles in the blood. This type of anemia can occur in sprue, pellagra, pregnancy, and in gastric and intestinal disorders.

Folic acid occurs in many plant and animal tissues, especially in the foliage of plants, from which it was named. Yeast, soybeans, wheat, liver, kidney, and eggs are good sources of this vitamin.

To function as a coenzyme, folic acid must be reduced to either **dihydrofolic acid** or **tetrahydrofolic acid.** The enzymes folic reductase and dihydrofolic reductase convert the vitamin to the active coenzyme tetrahydrofolic acid. The major role of tetrahydrofolic acid is as a carrier of one-carbon or formate units in the biosynthesis of purines, serine, glycine, and methyl groups.

Vitamin B_{12}

Vitamin B_{12} has a complex chemical structure that is centered about an atom of cobalt bound to the four nitrogen atoms of a corrin ring system, to a nucleotide, and to a cyanide group. The corrin ring system resembles the porphyrin ring system of hemoglobin in that it contains four pyrrole rings; however, in the corrin ring two of the pyrrole rings are joined directly rather than through a methene bridge. It is called **cyanocobalamin,** and is represented as follows:

Vitamin B_{12} (cyanocobalamin)

Vitamin B_{12}, like folic acid, is useful in the treatment of the anemias that develop in humans and animals. **Pernicious anemia** in particular responds most readily to treatment with the vitamin. In addition to increasing the hemoglobin and the red cell count, vitamin B_{12} administration also produces a remission of the neurological symptoms of anemia.

The best source of vitamin B_{12} is liver. Other sources include milk, beef extract, and culture media of microorganisms. Liver extracts also contain a hydroxycobalamin on which the cyanide group is replaced by a hydroxyl group. The coenzyme form of the vitamin occurs in nature and is known as coenzyme B_{12}. It is an unstable compound in which the CN or OH group attached to the cobalt atom in vitamin B_{12} is replaced by the nucleoside, adenosine, as shown:

Coenzyme B_{12}

The coenzyme is readily converted into either cyano- or hydroxycobalamin in the presence of cyanide or light. Coenzyme B_{12} functions in several important reactions in metabolism. It is involved in the isomerization of dicarboxylic acids; for example, it catalyzes the conversion of glutamic acid to methyl aspartic acid. The coenzyme also assists in the conversion of glycols and glycerol to aldehydes, in the biosynthesis of methyl groups, and in the synthesis of nucleosides.

Ascorbic Acid (Vitamin C)

Ascorbic acid is an enediol of a hexose sugar acid. The reduced or enediol form is readily oxidized to form **dehydroascorbic acid.** Both forms are biologically active; however, treatment with a weak alkali opens the oxide ring and produces an inactive molecule.

L-Ascorbic acid   Dehydroascorbic acid

A deficiency of ascorbic acid in the diet results in the disease known as **scurvy.** As early as 1720, citrus fruits were used as a cure for scurvy. The fact that all British ships were later required to carry stores of lime juice to prevent scurvy on long voyages led to the use of the terms "limey" for sailors, "lime juicers" for ships, and "lime house district" for the section of town in which sailors lived. Early symptoms of scurvy are loss of weight, anemia, and fatigue. As the disease progresses, the gums become swollen and bleed readily, and the teeth loosen. The bones become brittle and hemorrhages develop under the skin and in the mucous membrane. Acute scurvy is not commonly seen today, although many cases of subacute, or latent, scurvy are recognized. Symptoms such as sore receding gums, sores in the mouth, tendency to fatigue, lack of resistance to infections, defective teeth, and pains in the joints are indicative of **subacute scurvy.**

Man, monkeys, and guinea pigs are the only species that are known to be susceptible to the lack of ascorbic acid. Other animals possess the ability to synthesize this vitamin. The richest sources of ascorbic acid are citrus fruits, tomatoes, and green leafy vegetables. A large percentage of the ascorbic acid in foods is destroyed or lost in cooking. Prolonged boiling and the addition of sodium bicarbonate to maintain the green color of vegetables can destroy 70 to 90 per cent of the vitamin C content.

Ascorbic acid may function in oxidation or reduction processes in the body since it is a powerful reducing agent. The adrenal cortex contains appreciable amounts of ascorbic acid, which may function in the synthesis of steroid hormones in the adrenal gland. Ascorbic acid is also thought to be involved in hydroxylation reactions and in electron transport in the microsomal region of the cell. The biochemistry of ascorbic acid deficiency in the body is not as yet well understood. Linus Pauling (p. 23) has suggested that the common cold can be prevented or treated with large doses of ascorbic acid. Considerable controversy has arisen regarding this suggestion since neither the cause of the common cold nor the action of ascorbic acid in the body is clearly understood.

FAT-SOLUBLE VITAMINS

After vitamin C deficiency was related to **scurvy** and vitamin B to **beriberi,** it was noted in experiments prior to 1920 that certain animal fats such as butter and cod liver oil were capable of promoting growth in young rats which were fed a purified diet. These fat-soluble vitamins were first collectively called vitamin A, but now include vitamins A, D, E, and K.

Vitamin A

Vitamin A is closely related to the carotenoid pigments, alpha, beta, and gamma carotene and cryptoxanthin, which are polyunsaturated hydrocarbons. The carotene pigment, beta carotene, has an all-*trans* structure and is an active precursor of the vitamin.

β-Carotene (all-*trans*)

Vitamin A represents half the beta carotene molecule with the ends oxidized to primary alcohol groups.

Vitamin A (all-*trans*)

The carotene pigments and cryptoxanthine can be converted into vitamin A in the animal body. The vitamin is soluble in fat and fat solvents and is a liquid at room temperature.

A diet deficient in vitamin A will not support growth, and the deficiency adversely affects the epithelial cells of the mucous membrane of the eye, the respiratory tract, and the genitourinary tract. The process in which the mucous membrane hardens and dries up is known as **keratinization.** The eye is first to show the effect of a deficiency and one of the first symptoms of the lack of the vitamin is **night blindness.** Later the eyes develop a disease called **xerophthalmia.** This disease is characterized by inflamed eyes and eyelids. The eyes ultimately become infected, and when this infection involves the cornea and lens, sight is permanently lost. A continued deficiency of vitamin A results in extensive infection in the respiratory tract, the digestive tract, and the urinary tract. Vitamin A deficiency also causes sterility, since it affects the lining of the genital tract. It is therefore necessary for normal reproduction and lactation. More recently the vitamin has been found essential for the synthesis of mucopolysaccharides which form the ground substance of structural tissue and for the maintenance of the stability of cellular membranes and the membranes of subcellular particles such as lysosomes and mitochondria.

Fish liver oils are potent sources of vitamin A. Eggs, liver, milk and dairy products, green vegetables, and tomatoes are good food sources of the vitamin. The body has the ability to store vitamin A in the liver when it is present in the food in excess of the body requirements. Infants obtain a store of the vitamin in the first milk (colostrum) of the mother, which is ten to one hundred times as rich in vitamin A as ordinary milk.

THE BIOCHEMISTRY OF VISION

The visual process in the eyes of man and animals involves two types of photoreceptors: rods for vision in dim light, and cones for vision in bright light. Little is known of the mechanism involving the visual process in the cones. We owe to George Wald's extensive research on vision our more complete knowledge of the visual mechanism in the rods.

Vitamin A is involved in the visual process in the formation of the visual pigment **rhodopsin,** which is a complex composed of retinal (formerly called retinine) and opsin (a protein). **Retinal** has been identified as vitamin A aldehyde; it may exist in the *cis* or all-*trans* form. All-*trans*-retinal is vitamin A with the primary alcohol group oxidized to an aldehyde. The structure of Δ^{11} *cis*-retinal is shown as follows:

Δ^{11} *cis*-retinal

The relation between rhodopsin, retinal, and vitamin A and the visual cycle is shown on p. 100.

When light strikes rhodopsin, isomerization of Δ^{11} *cis*-retinal to the all-*trans*-retinal occurs, and the complex splits into the protein opsin and all-*trans*-retinal. The latter compound is reduced to all-*trans*-vitamin A. In the regeneration of the visual pigment rhodopsin, the all-*trans*-vitamin A is first isomerized to Δ^{11} *cis*-vitamin A, then oxidized to Δ^{11} *cis*-retinal. There is a loss of vitamin A during the regeneration of rhodopsin after exposure to light. Since this supply of vitamin A must come from the blood, a normal rate of rhodopsin synthesis is therefore dependent on the vitamin A concentration in the blood. In the vitamin-conscious culture of the United States, most of our citizens have a more than adequate dietary intake of vitamin A. Dietary deficiencies that exist in underdeveloped countries result in many cases of xerophthalmia.

Vitamin D

Chemical Nature. Several compounds with vitamin D activity exist, although only two of them commonly occur in antirachitic drugs and foods. These two compounds are produced by the irradiation of ergosterol and 7-dehydrocholesterol with ultraviolet light. Ergosterol is a sterol that occurs in yeast and molds, whereas 7-dehydrocholesterol is found in the skin of animals. Irradiated ergosterol is called **calciferol,** or vitamin D_2; irradiated 7-dehydrocholesterol is called vitamin D_3. The structures of the two forms of vitamin D are very similar.

The lack of vitamin D in the diet of infants and children results in an abnormal formation of the bones, a disease called **rickets.** Calcium, phosphorus, and vitamin D are all involved in the formation of bones and teeth. Bowed legs, a "rachitic rosary" of the ribs, an abnormal formation of the ribs known as "pigeon breast," and poor tooth development are common signs of vitamin D deficiency in small children. Rickets does not occur in adults after bone formation is complete, although the condition of **osteo-malacia** may occur in women after several pregnancies. In osteomalacia, the bones soften and abnormalities of the bony structure may occur.

The fish liver oils are the most potent sources of vitamin D; fish such as sardines, salmon, and herring are the richest food sources. The ultraviolet rays in sunlight form vitamin D by irradiation of 7-dehydrocholesterol in the skin. Children who play outdoors in the summer materially increase the vitamin D content of their bodies. Milk in particular and other foods have their vitamin D content increased by the addition of small amounts of irradiated ergosterol.

The main function of vitamin D in the body is to increase the utilization of calcium and phosphorus in the normal formation of bones and teeth. The exact mode of action of the vitamin is not known, although it increases calcium and phosphorus absorption from the intestine, stimulates the activity of the enzyme phosphatase, and is essential for normal growth.

Vitamin E

Vitamin E is chemically related to a group of compounds called **tocopherols**. Alpha-, beta-, and gamma-tocopherol have vitamin E activity, but alpha-tocopherol is the most potent.

Alpha-tocopherol

The other two tocopherols differ only in the number and position of the CH_3 groups on the aromatic ring. Beta is a 1,4-di-CH_3, and gamma a 1,2-di-CH_3 derivative. Vitamin E is stable to heat but is destroyed by oxidizing agents and ultraviolet light. Oxidative rancidity of fats rapidly destroys the potency of the vitamin.

It has been known for several years that when animals such as rabbits and rats are maintained on a vitamin E deficient diet, muscular dystrophy, creatinuria, and anemia develop in rabbits, and changes in reproductive organs and function occur in rats. The richest source of vitamin E is wheat germ oil. Corn oil, cottonseed oil, egg yolk, meat, and green leafy vegetables are good sources of this vitamin.

The tocopherols are excellent antioxidants and prevent the oxidation of several substances in the body, including unsaturated fatty acids and vitamin A. As an antioxidant, vitamin E may protect mitochondrial systems in the cell from irreversible oxidation by lipid peroxides. It may also protect lung tissue from damage by oxidants present in smog-contaminated atmospheres.

Vitamin K

Vitamin K is a derivative of 1,4-naphthoquinone, as is illustrated in the formula for vitamin K:

Vitamin K

The 2-methyl-1,4-napthoquinones and naphthohydroquinones possess vitamin activity. Vitamin K is fat-soluble and therefore is soluble in ordinary fat solvents. It is stable to heat, but is destroyed by alkalies, acids, oxidizing agents, and light.

A diet lacking in vitamin K will cause an increase in the clotting time of blood. This condition produces hemorrhages under the skin and in the muscle tissue. The abnormality in the clotting mechanism is due to a reduction in the formation of **prothrombin,** one of the factors in the normal process.

Rich sources of vitamin K_1 are alfalfa, spinach, cabbage, and kale. The vitamin K_2 series is present in bacteria. The bacteria present in putrefying fish meal are capable of synthesizing vitamin K_2 and are potent sources.

The vitamin is essential in the synthesis of prothrombin by the liver, but the exact mechanism of synthesis is as yet unknown. It is also thought to be involved in the metabolic reactions and electron transport systems in the mitochondria of cells. The role of **prothrombin** in the clotting of blood is shown in the following sequence:

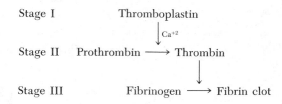

Although the clotting process is represented as two simple equations, it involves a large series of factors and reactions in which prothrombin is converted to the enzyme thrombin, which in turn converts plasma **fibrinogen** to the **fibrin clot.**

IMPORTANT TERMS AND CONCEPTS

calciferol
Δ^{11} *cis*-retinal
coenzyme A
cyanocobalamin
flavin adenine dinucleotide

nicotinamide adenine dinucleotide
pyridoxal phosphate
tetrahydrofolic acid
thiamine pyrophosphate
tocopherol

QUESTIONS

1. Describe the symptoms of a common deficiency disease caused by the lack of a vitamin.

2. What is the difference between a coenzyme and a prosthetic group? Why is an apoenzyme inactive?

3. What type of vitamin is often found as a part of the coenzyme molecule? Explain.

4. Name the important coenzymes that are involved in oxidation-reduction reactions in the body.

5. Write the structure for acetyl coenzyme A, pointing out its component parts.

6. Why are adequate amounts of vitamin C in the diet important in nutrition?

7. What is the relation between Δ^{11} *cis*-retinal, vitamin A, and vision?

8. Why are babies and young children often given doses of cod liver oil?

9. Discuss the function of tocopherols in the body and in food products.

Chapter 8

BIOCHEMICAL ENERGY

The *objectives* of this chapter are to enable the student to:

1. Recognize the relation between ΔG and $\Delta G°$.
2. Calculate the $\Delta G°$ for a coupled reaction that utilizes ATP as the driving force.
3. Recognize ATP as a high-energy compound and its relation to ADP and AMP.
4. Illustrate the transfer of high-energy phosphate groups to ADP to form ATP.
5. Explain the process of oxidative phosphorylation and its site of action in the mitochondria.
6. Recognize the difference in moles of ATP formed by NADH and FADH in the electron transport system.

We commonly think of energy in terms of heat or electricity to heat our homes, run electric motors, or operate electric appliances. While the body may produce heat by muscular work, the living cell operates under essentially isothermal conditions, without differences in pressure and at a nearly neutral pH. The energy that cells produce in their reactions is transformed into chemical energy, which is then used in the mechanical work of muscular contraction, the osmotic work required to transport materials in and out of the cell, and the chemical work involved in the synthesis of cell components. All of the reactions involved in these processes take place at a virtually constant temperature. A portion of the chemical energy that is formed in the cell is stored in high-energy compounds. These high-energy compounds are valuable since they are used to drive essential reactions in the metabolic cycles of carbohydrate, lipid, and protein metabolism.

FREE ENERGY CHANGE IN REACTIONS

The useful form of energy that is taken in by cells is called **free energy,** which may be defined as that type of energy which can do work at constant temperature and pressure. In a cellular reaction involving reactants A and products B, an equation that expresses the relation between the change in free energy, ΔG, and a change in concentration of products and reactants is as follows:

$$\Delta G = \Delta G° + RT\ 2.3 \log_{10} \frac{[B]}{[A]}$$

where $\Delta G°$ is the **standard free-energy change** of the reaction, R is the gas constant, T is the absolute temperature $(273 + °C)$, and the activities of B and A are expressed as concentrations in moles per liter. At equilibrium, $\Delta G = 0$ and the ratio of [B] to [A] equals K. By substituting in the previous equation

$$0 = \Delta G° + RT\ 2.3 \log_{10} K,\ or$$
$$\Delta G° = -RT\ 2.3 \log_{10} K$$

From these equations we could calculate $\Delta G°$ knowing the equilibrium constant K or the concentration of the products and reactants at equilibrium. Once $\Delta G°$ is known, we can calculate ΔG or the free-energy change of any reaction. The ΔG of a reaction depends on the standard free-energy change and on the concentrations of reactants and products. For example, if, when equilibrium is reached in the enzymatic hydrolysis of adenosine monophosphate at 25°C to form adenosine plus phosphoric acid (Ⓟ), the concentration of AMP is 0.002M and the concentration of the products is 0.120M, we may calculate $\Delta G°$ as follows:

$$K = \frac{B}{A} = \frac{A + Ⓟ}{AMP} = \frac{0.120}{0.002} = 60$$
$$\Delta G° = -RT\ 2.303 \log_{10} K = -1.987 \times 298 \times 2.303 \log_{10} 60$$
$$\Delta G° = -1363 \log_{10} 60 = -1363 \times 1.78 = -2.40\ kcal$$

The standard free-energy change, $\Delta G°$, of a reaction is theoretically measured under constant conditions of pH (7.0), temperature (25°C), and concentration. This would involve, for example, starting a reaction with 1 mole of adenosine monophosphate and converting it to 1 mole of adenosine under conditions in which the concentration of each compound is maintained at 1 molar. This is obviously a situation different from the experimental conditions in the above example, and would be difficult to maintain in a test tube or in a cell. Since the heterogeneous systems in the cell do not achieve the theoretical conditions, the $\Delta G°$ values in this chapter are at best comparative approximations, but the concept is valuable in intermediary metabolism.

COUPLED REACTIONS

There are many chemical reactions in a cell that have a $+\Delta G°$ and will not proceed in a left to right direction without assistance. Removal of one or more of the products of the reaction may force it to the right, or it may be coupled to a highly exergonic reaction. In general, an endergonic reaction may be coupled with an exergonic reaction so that energy is delivered to the endergonic reaction. In such coupled reactions the algebraic sum of the free-energy changes in the two reactions must be negative in sign (a net decline in free energy) for the coupled reaction to occur. The energy in ATP (adenosine triphosphate), as represented by a $-\Delta G°$ of 7.3 kcal when ATP is converted to ADP (adenosine diphosphate), is often used to drive endergonic reactions. In the formation of glucose-6-phosphate from glucose and phosphate, about 3.3 kcal are required. If this reaction is coupled with the ATP $\rightarrow$ ADP reaction in the presence of the enzyme hexokinase, the following $\Delta G°$ results:

$$Glucose + ATP \xrightarrow{\text{hexokinase}} Glucose\text{-}6\text{-}PO_4 + ADP$$
$$\Delta G° = -7.3 + 3.3 = -4.0\ kcal\ (approximately)$$

Many **coupled reactions** in the cell involve the formation of a common intermediate with the assistance of ATP. The formation of sucrose from glucose and fructose, for example, has a $\Delta G°$ of $+5.5$ kcal and requires the conversion of ATP to ADP to drive the reaction to completion. The coupled reaction and formation of a common intermediate, glucose-1-PO_4, can be represented as follows:

$$\text{Glucose} + \text{ATP} \rightarrow \text{Glucose-1-}PO_4 + \text{ADP}$$
$$\text{Glucose-1-}PO_4 + \text{Fructose} \rightarrow \text{Sucrose} + P_i$$
$$\Delta G° = -7.3 + 5.5 = -1.8 \text{ kcal (approximately)}$$

In the first reaction, the terminal PO_4 group of ATP was transferred to glucose and with it some of the energy of the ATP. In the second reaction, the energy-enriched glucose-1-PO_4 reacts with fructose to form sucrose.

HIGH-ENERGY COMPOUNDS

In the preceding section we learned that many endergonic reactions in a cell can be coupled to an exergonic reaction to obtain the energy to drive the cellular reaction to the right. Early investigations on the nature of muscular contraction revealed that the presence of the high-energy compound creatine phosphate is a driving force in muscle reactions. Studies on the oxidation of glucose and especially the metabolic cycles of carbohydrate oxidation emphasized the role of **adenosine triphosphate, ATP,** and this energy-rich compound has become the key in linking endergonic processes to those that are exergonic.

High-energy compounds are often complex phosphate esters that yield large amounts of free energy on hydrolysis. A more detailed consideration of the energy released by the stepwise hydrolysis of ATP will illustrate the high-energy concept.

Adenosine triphosphate, ATP

$$\text{ATP} \xrightarrow{\text{hydrolysis}} \text{ADP} + H_3PO_4 \qquad \Delta G° = -7.3 \text{ kcal}$$
$$\text{ADP} \xrightarrow{\text{hydrolysis}} \text{AMP} + H_3PO_4 \qquad \Delta G° = -7.3 \text{ kcal}$$
$$\text{AMP} \xrightarrow{\text{hydrolysis}} \text{Adenosine} + H_3PO_4 \qquad \Delta G° = -2.2 \text{ kcal}$$

Several explanations have been proposed for the release of energy on the hydrolysis of high-energy compounds. These include the fact that these compounds are unstable in acid and alkaline solutions and are readily hydrolyzed. Also, the hydrolysis products, inorganic phosphate, ADP, and AMP, have many more resonance possibilities than the parent ATP. A major reason for the release of energy involves the *type of bond structure* in these compounds. The β and γ bonds in ATP are anhydride linkages that have closely spaced negative charges, which repel each other strongly. Some of this electrical stress is relieved on hydrolysis.

TOPIC OF CURRENT INTEREST

THE WORLD OF ATP

Adenosine triphosphate, or ATP, was first discovered in 1929 in muscle extracts, and was thought to be primarily concerned with muscular contraction. In the 1930s two well-known German biochemists, Otto Warburg and Otto Meyerhof, demonstrated that ATP is generated from ADP in the anaerobic pathway of glucose to lactic acid in muscle. It was later shown that ATP is also generated from ADP during aerobic oxidations in animal tissue. Also in the 1930s it was discovered that ATP is hydrolyzed to ADP by myosin, the major contractile protein of muscle, and that ATP is required to phosphorylate glucose and thus prepare it for the biosynthesis of glycogen.

These observations and many others were combined by Lipmann in 1941 into a general hypothesis for energy transfer in living cells. He postulated that ATP functions in a cyclic manner as the carrier of chemical energy from the catabolic reactions of metabolism, which yield energy, to the various cellular processes that require energy input. The ATP molecules generated in these reactions were thought to donate their terminal phosphate groups to specific acceptor molecules, energizing them for carrying out various energy-requiring functions in the cell. These functions include the contraction of muscles, the synthesis of cellular proteins and other macromolecules, and the active transport of cell nutrients and inorganic ions across membranes against gradients of concentration. As ATP delivers energy to these processes, the ATP is cleaved to ADP and inorganic phosphate. The ADP molecule is then rephosphorylated by the energy-yielding oxidation of fuels to produce ATP again, thus completing the cellular energy cycle.

In 1930 Lohmann proposed the structure of ATP, which was finally confirmed through chemical synthesis by Todd and co-workers in 1948. In the plant and animal world ATP, ADP, and AMP are widely distributed; for example, in the aqueous phase of living cells the sum of their concentrations varies between 2 and 10 mM. At pH 7.0 both ATP and ADP are highly ionized anions, with 4 ionizable protons on the phosphate groups of ATP and 3 ionizable protons on ADP. In the intracellular fluid there exists a relatively high concentration of Mg^{2+}, which readily combines with the negative charges on the triphosphate group of ATP to form $MgATP^{2-}$. In the majority of the enzyme reactions in the cell in which ATP is involved, the active form of the compound is the $MgATP^{2-}$ complex.

When ATP is hydrolyzed to form ADP, a pyrophosphate bond, the oxygen bridge between two phosphorus atoms, is cleaved. The hydrolysis of a pyrophosphate bond is accompanied by a highly negative standard free energy change. These energy-rich or high-energy bonds are sometimes designated by a special symbol. ATP could be written as A—P$\sim$P$\sim$P, since the last two bonds are energy-rich anhydride bonds compared to the common ester bond between adenine and the first phosphorus atom. ATP is involved in so many metabolic reactions in the cell that it serves as a common currency for exchange of energy among reactions. It is formed by highly exergonic reactions and is hydrolyzed to drive endergonic processes in the cell. In addition, it serves a very important function in the regeneration of other nucleoside triphosphates, which are energy-rich compounds used in synthetic cellular processes and in driving other reactions. When cytidine triphosphate (CTP), uridine triphosphate (UTP), and guanosine triphosphate (GTP) are used in cellular synthesis or in driving reactions, they release energy and are hydrolyzed to diphosphates or monophosphates. ATP and two enzymes, nucleoside diphosphate kinase and nucleoside monophosphate kinase, are required to transfer phosphate groups from ATP to re-form CTP, UTP, and GTP. The interchange between these high-energy compounds is illustrated by the nucleotide phosphate pool, which points up in very definite fashion the importance of ATP in the cell.

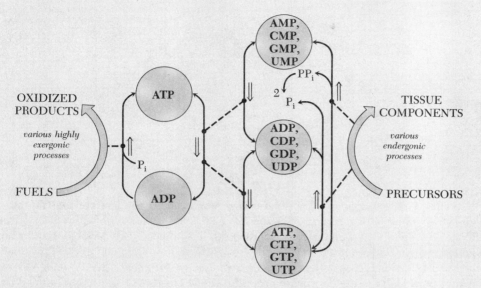

(From McGilvery: Biochemical Concepts, Philadelphia, W. B. Saunders Company, 1975, p. 147.)

Other phosphorus-containing, high-energy compounds include:

1,3-Diphosphoglyceric acid
$\Delta G = -11.8$ kcal

Phosphoenolpyruvic acid
$\Delta G° = -14.8$ kcal

Creatine phosphate
$\Delta G° = -10.0$ kcal

Acetyl coenzyme A
$\Delta G° = -8.2$ kcal

The top two compounds in the previous illustrations have anhydride linkages between a phosphate and either a carbonyl or acid enol group. Creatine phosphate, the high-energy compound in muscle, has a direct linkage between phosphate and nitrogen, whereas the acyl mercaptide linkage in acetyl coenzyme A is also characteristic of an energy-rich compound. In every instance the high-energy compound is readily hydrolyzed to products that undergo spontaneous reactions. These reactions result in forms that are thermodynamically more stable.

Simple phosphate esters, such as AMP, glucose-6-phosphate, and 3-phosphoglyceric acid, are not considered as high-energy compounds, and yield less energy on hydrolysis.

Glucose-6-phosphate
$\Delta G° = -3.3$ kcal

3-Phosphoglyceric acid
$\Delta G° = -2.2$ kcal

Some of the compounds described above may be used to illustrate the transfer of high-energy phosphate groups to ADP to form ATP, and the transfer of these groups from ATP to low-energy phosphate compounds. Both 1,3-diphosphoglyceric acid and phosphoenolpyruvic acid are important metabolites formed during the energy-yielding anaerobic breakdown of glucose in the cell. They have much more negative $\Delta G°$s than ATP and can transfer a high-energy phosphate group to ADP to form ATP (p. 109). The ATP thus formed or present in the cell has a much greater negative $\Delta G°$ than glycerol or glucose and can transfer a high-energy phosphate group to these compounds, forming 3-phosphoglyceric acid or glucose-6-phosphate. These transfer reactions illustrate the flow of phosphate groups from high-energy phosphate donors to low-energy acceptors by way of the ADP-ATP system. This emphasizes the role of the ADP-ATP system as an obligatory common intermediate which carries phosphate groups from high-energy phosphate compounds generated during catabolism to low-energy compounds, which thus become energized.

THE FORMATION OF ATP

Since adenosine triphosphate has been marked as a key compound in the storage of chemical energy and in the coupling of exergonic reactions to endergonic reactions in the cell, it is a major driving force in the metabolic reactions in the tissue. Although ATP can be formed by light energy in the process of photosynthesis, which will be discussed in Chapter 10, the present discussion will consider its formation in the cytoplasm in the absence of oxygen (substrate level phosphorylation) and in the **mitochondria** by the process of oxidative phosphorylation.

Substrate Level ATP

In the anaerobic scheme of carbohydrate metabolism (Embden-Meyerhof pathway) glucose is phosphorylated and is eventually broken down to the 3-carbon phosphorylated derivative. In the following two reactions ADP is converted into the energy-rich compound ATP with the assistance of catalysts, called **enzymes.** These reactions can take place in the absence of O_2 and in the cytoplasm, and are termed **substrate level phosphorylations.**

$$
\begin{array}{c}
\overset{\displaystyle O}{\overset{\|}{C}}-O-\overset{\displaystyle O}{\overset{\|}{P}}-OH \\
| \qquad\quad | \\
\qquad\quad O^- \\
H-\overset{|}{C}-OH \\
| \qquad\quad O \\
H_2\overset{|}{C}-O-\overset{\|}{P}-OH \\
\qquad\quad | \\
\qquad\quad O^-
\end{array}
\;+\; ADP \xrightarrow[\text{kinase}]{\text{phosphoglycero-}}\;
\begin{array}{c}
COOH \\
| \\
H-\overset{|}{C}-OH \\
| \qquad\quad O \\
H_2\overset{|}{C}-O-\overset{\|}{P}-OH \\
\qquad\quad | \\
\qquad\quad O^-
\end{array}
\;+\; ATP
$$

1,3-Diphosphoglyceric   3-Phosphoglyceric
    acid   acid

$$
\begin{array}{c}
COOH \\
| \qquad\quad O \\
\overset{\displaystyle }{C}-O-\overset{\|}{P}-OH \\
\| \qquad\quad | \\
\qquad\quad O^- \\
CH_2
\end{array}
\;+\; ADP \xrightarrow[\text{kinase}]{\text{pyruvic}}\;
\begin{array}{c}
COOH \\
| \\
C=O \\
| \\
CH_3
\end{array}
\;+\; ATP
$$

Phosphoenolpyruvic   Pyruvic acid
    acid

Oxidative Phosphorylation ATP

One of the major aerobic or oxidative schemes of carbohydrate metabolism (Krebs cycle) involves the reaction of intermediate compounds with the production of several moles of ATP. An **electron transport system** in the mitochondria of the cell actively transports electrons from a reduced metabolite to oxygen with the assistance of enzymes and coenzymes, as shown in Figure 8–1. P_i is inorganic phosphate, NAD is nicotinamide adenine dinucleotide, FMN is flavin mononucleotide, and FAD is flavin adenine dinucleotide; these three coenzymes were discussed in the preceding chapter. Other intermediate compounds in Figure 8–1 between the reduced metabolite and oxygen, beside NAD and FAD, are coenzyme Q and the cytochromes. The overall reactions involve first

$$NAD + Metabolite \cdot H_2 \rightarrow Metabolite + NADH + H^+$$

then

$$NADH + 3\,ADP + 3\,P_i + \tfrac{1}{2}O_2 \rightarrow NAD + 3\,ATP + H_2O$$

As may be seen from Figure 8–1, ATP is generated at three sites on the electron transport chain with the following electron transfers: from NADH to coenzyme Q through flavoprotein FMN, from coenzyme Q to cytochrome c through cytochrome b, and from cytochrome c to O_2 through cytochrome a_3. Three moles of ATP may therefore be generated by the passage of electrons from a mole of substrate through NAD to molecular oxygen, but only two moles may be generated if electrons are transferred directly from the substrate to coenzyme Q through FAD because this transfer bypasses the first phosphorylation site.

The enzymes of electron transport are located on the inner membrane of the mitochondria. This membrane is thought to consist of repeating units, each composed of a headpiece which projects into the matrix, joined by a stalk to a basepiece (Fig. 8–2). The basic electron transport chain is located in the basepieces. Each basepiece is considered a complex containing a portion of the electron transport enzymes. Four such complexes plus NADH, coenzyme Q, and cytochome c constitute a complete electron transport system. The energy generated by electron transport in the basepiece complex is transmitted through the stalk protein to the headpiece, where it is converted into the high-energy bond of ATP.

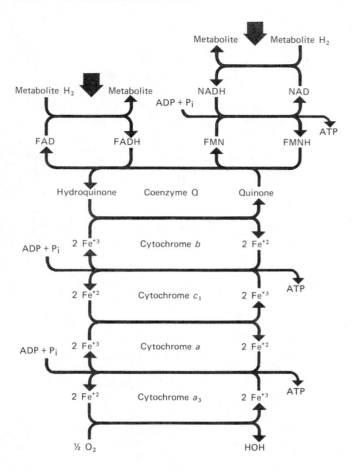

FIGURE 8-1 Oxidative phosphorylation and the electron transport system.

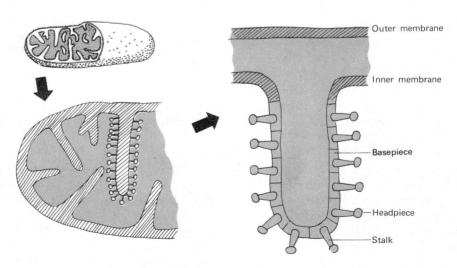

FIGURE 8-2 Proposed structure of the inner membrane of mitochondria.

NAD, **nicotinamide adenine dinucleotide** (structure shown in the preceding chapter), is a dinucleotide composed of AMP linked to nicotinamide-ribose-phosphate. The nicotinamide portion of the molecule is involved in the oxidation and reduction reactions in oxidative phosphorylation. The remaining portion of the molecule is represented as Ribose-Ⓟ-O-Ⓟ-adenosine, where Ⓟ stands for phosphoric acid. The reduction of NAD to form NADH may be represented as:

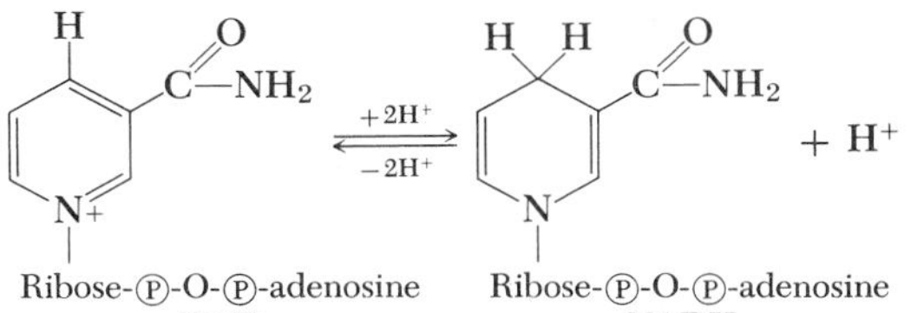

NAD NADH

FAD, **flavin adenine dinucleotide** (structure shown in preceding chapter), is a dinucleotide composed of a flavin-ribose-phosphate linked to AMP. The reduction of FAD in the electron transport system may be represented as follows:

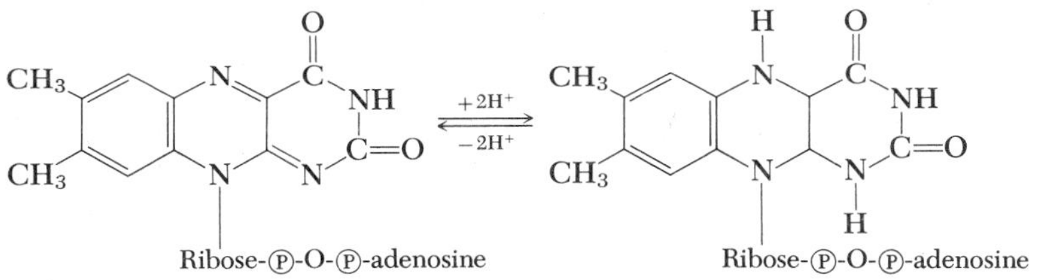

FAD FADH

Coenzyme Q is a lipid-soluble quinone, sometimes called ubiquinone-10 for the ten isoprene units found in the side chain (the number may vary from 0 to 10). This coenzyme is readily reduced to the hydroquinone form during the transport of hydrogen, as shown:

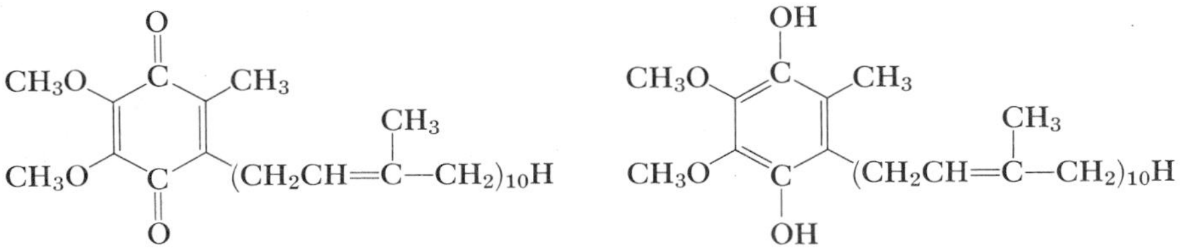

Coenzyme Q (quinone form)

Coenzyme Q (hydroquinone form)

The **cytochromes** are oxidation-reduction pigments that consist of iron-porphyrin complexes known as **heme,** which is also an integral part of hemoglobin, the respiratory pigment of the red blood cells. The heme in cytochrome c, for example, is attached to a protein molecule by coordination with two basic amino acid residues, and by thioether linkages formed by the addition of a sulfhydryl group from each of two molecules of cysteine in the protein molecule. Cytochrome c is an electron carrier in the oxidative phosphorylation cycle, in which the iron atom of heme is changed from Fe^{+++} to Fe^{++} as shown on the next page.

Protein

HCCH₃ H   CH₃
C

CH₃—

—CH₂CH₂COOH

HC
CH₃

Fe⁺⁺⁺   CH

—S—C—
H

—CH₂CH₂COOH

CH₃ H CH₃
C
N

Cytochrome c
(oxidized)

Protein

HCCH₃ H   CH₃
C

CH₃—

—CH₂CH₂COOH

HC
CH₃

Fe⁺⁺   CH

—S—C—
H

—CH₂CH₂COOH

CH₃ H CH₃
C
N

Cytochrome c
(reduced)

The exact nature and detailed functional mechanism of the oxidative phosphorylation cycle is as yet not completely understood. Recently Green and his co-workers have isolated a large **electron transport particle** from the mitochondria of cells. They then separated the particle into four complexes, and, after a study of their composition and function, proposed the following relationship between the complexes and the electron transport system:

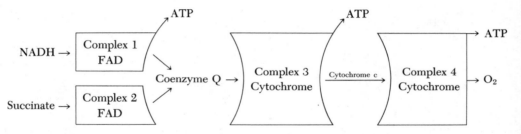

Biochemical energy in the form of ATP is an essential driving force in many metabolic reactions in the cells and tissues. As has been described, several complex reactions are involved in the synthesis of this vital compound, and it should be emphasized that three moles of ATP are formed when the electrons from NADH are transported through the system to oxygen. Also, two moles of ATP are formed when electrons from FADH are transported to oxygen. These relationships will assist in the understanding of the energy balance in the metabolic cycles.

IMPORTANT TERMS AND CONCEPTS

ATP
coenzyme Q
coupled reactions
cytochromes
electron transport system

FADH
free-energy change
high-energy compounds
NADH
oxidative phosphorylation
standard free-energy change

QUESTIONS

1. Discuss the limitations of energy production in the living cell.

2. What is the relation between the standard free-energy change of a reaction, $\Delta G°$, and the free-energy change, ΔG?

3. In the conversion of glucose-1-phosphate to glucose-6-phosphate, an equilibrium concentration of 0.001 M glucose-1-PO_4 and 0.022 M glucose-6-PO_4 is obtained. Calculate the $\Delta G°$ of the reaction.

4. The formation of an ester from an acid and an alcohol resulted in a $\Delta G°$ of $+2.0$ kcal. ATP formed an intermediate with the acid to drive the reaction to completion. Represent the reaction and calculate the new $\Delta G°$.

5. Discuss the functions of ATP in the body.

6. Briefly explain why creatine phosphate is a high-energy compound.

7. If 3-phosphoglyceric acid were converted to 1,3-diphosphoglyceric acid, what would happen to the ATP and what energy change would you expect?

8. Why is the oxidative phosphorylation mechanism also called the electron transport system? Explain.

9. Explain why only two moles of ATP are formed in the electron transport system when electrons from FADH are transported to oxygen.

10. Trace the process of the electron transport system from the basepiece to the headpiece on the inner membrane of the mitochondria.

Chapter 9

INTRODUCTION TO METABOLISM

The *objectives* of this chapter are to enable the student to:

1. Describe the processes of digestion and absorption of carbohydrates.
2. Describe the processes of digestion and absorption of dietary fat.
3. Describe the processes of digestion and absorption of dietary protein.
4. Define the processes of intermediary metabolism in the cell.
5. Distinguish between the processes of catabolism and anabolism in the outline of intermediary metabolism.

A simple definition of **metabolism** would be the consideration of all the enzymatic reactions occurring in the living cell. Since we are obviously not aware of all the reactions that occur in the cell, we may begin by breaking metabolism into several component parts. Two major divisions are **anabolism,** or the biosynthetic processes, and **catabolism,** or the biodegradative processes that result in metabolites needed by the cell and the chemical energy used in cellular reactions. Prior to the metabolic reactions in the cell, the major foodstuffs of the diet must undergo the processes of digestion and absorption.

SALIVARY DIGESTION

Food taken into the mouth is broken into smaller pieces by chewing and is mixed with saliva, the first of the digestive fluids. Saliva contains **mucin,** a glycoprotein that makes the saliva slippery, and **ptyalin,** an enzyme that catalyzes the hydrolysis of starch to maltose. Since this enzyme has little time to act on starches in the mouth, its main activity takes place in the stomach before it is inactivated by the acid gastric contents. There are no enzymes in the saliva that act on dietary fats or proteins.

GASTRIC DIGESTION

When food is swallowed it passes through the esophagus into the stomach. During the process of digestion the food is mixed with gastric juice, which is secreted by small tubular glands located in the walls of the stomach. Gastric juice is a pale yellow, strongly acid solution containing the enzymes **pepsin** and **rennin.** Pepsin initiates the hydrolysis of large protein molecules into smaller, more soluble molecules of **proteoses** and **peptones,** whereas rennin converts casein of milk into a soluble protein. The optimum pH of pepsin is 1.5 to 2; thus, it is ideally suited for the digestion of protein in normal stomach contents, the pH of which is 1.6 to 1.8. The mixing action of the stomach musculature and the process of digestion produce a liquid mixture called chyme, which passes through the pyloric opening into the intestine.

INTESTINAL DIGESTION

The acid chyme is neutralized by the alkalinity of the three digestive fluids, **pancreatic juice, intestinal juice,** and **bile,** in the first part of the small intestine, the duodenum. When fat enters the duodenum, the gastrointestinal tract hormone **cholecystokinin** is secreted and is carried by the blood to the gallbladder, where it stimulates that organ to empty its bile into the small intestine. Bile acids and bile salts are good detergents and emulsify fats for digestion by **pancreatic lipase.** Another hormone that is secreted when the chyme enters the duodenum is **secretin.** This hormone enters the circulation and stimulates the pancreas to release pancreatic juice into the intestine. There are enzymes in the pancreatic juice that are capable of digesting proteins, fats, and carbohydrates. The pancreatic proteases are **trypsin, chymotrypsin,** and **carboxypolypeptidase,** whereas the pancreatic lipase is called **steapsin.** The enzyme **amylopsin** in pancreatic juice is an amylase similar to ptyalin in the saliva. Pancreatic lipase is activated by bile salts and splits fats into fatty acids, glycerol, soaps, mono- and diglycerides. Native proteins and the proteoses and peptones that result from the action of pepsin are gradually split into amino acids by the proteases of the pancreatic juice and **aminopolypeptidase** and **dipeptidase** in the intestinal juice. The intestinal juice also contains three disaccharide-splitting enzymes, **sucrase, lactase,** and **maltase.** Cane sugar is the main source of dietary sucrose; milk contains lactose; and maltose comes from the partial digestion of starch by ptyalin and amylopsin. Sucrase, lactase, and maltase split these disaccharides into their constituent monosaccharides, thus completing the digestion of carbohydrates.

ABSORPTION

The monosaccharides glucose, fructose, and galactose are absorbed directly into the bloodstream through the capillary blood vessels of the **villi.** The villi are fingerlike projections on the inner surface of the small intestine that greatly increase the effective absorbing surface. There are approximately five million villi in the human small intestine, and each villus is richly supplied with both lymph and blood vessels. Considerable evidence exists to indicate that the intestinal mucosa possesses the property of **selective absorption,** which is not possessed by a nonliving membrane. Enzyme mechanisms requiring energy are involved in the selective absorption of all molecules that pass through the intestinal mucosa. The rate of absorption is not determined by the size of the molecule but by the specific mechanism.

In the absorption process, as the end products of fat digestion pass through the villi of the intestinal mucosa they are reconverted into triglycerides, which then enter the lymph circulation. Bile salts are essential in absorption, both because of their effect on the solubility of fatty acids and because of their direct involvement in the absorption process.

Amino acids are absorbed through the intestinal mucosa directly into the bloodstream by an active process that requires energy and enzymes. After absorption, the amino acids are carried by the portal circulation to the liver and subsequently to all the tissues of the body.

INTERMEDIARY METABOLISM

Intermediary metabolism is concerned with the molecules presented to the cells and tissues following the process of absorption and involves a multitude of enzyme-catalyzed reactions in different parts of the cell. The specific details of carbohydrate, lipid, and protein metabolism will be presented in subsequent chapters; however, an overview of metabolism in the body as outlined in Figure 9–1 will assist in understanding the details.

Following absorption into the cell, the end products of digestion are further converted into simpler molecules by the process of catabolism. Hexoses, pentoses, and glycerol are converted into the 3-carbon phosphorylated sugar glyceraldehyde-3-phosphate and then into the 2-carbon acetyl group of acetyl coenzyme A. The fatty acids and amino acids from digestion and absorption are also broken down in the process of catabolism to **acetyl coenzyme A.** This common end product of catabolism, acetyl CoA, is then fed into the Krebs cycle, where the acetyl group is eventually catabolized into CO_2 with the production of chemical energy in the form of ATP. From the above discussion and Figure 9–1 it may be seen that the various pathways of catabolism flow toward a final common pathway, the Krebs cycle.

In Figure 9–1 it is apparent that the flow of anabolism is opposite that of catabolism. In the Krebs cycle α keto acids are formed, which may be aminated to form amino acids used in the synthesis of protein molecules. Acetyl groups from acetyl CoA may be assembled into fatty acids and eventually into lipids. Also, the acetyl groups from acetyl CoA may be converted back to pyruvate and proceed up the chain of reactions to the polysaccharide glycogen.

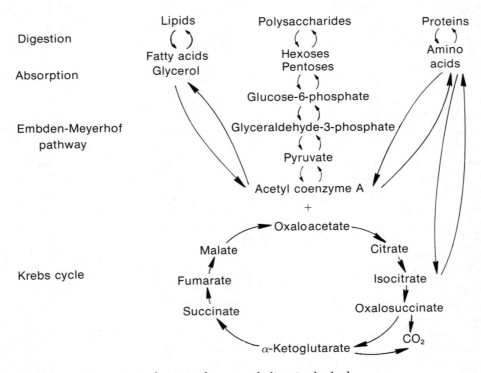

Figure 9–1   An overview of intermediary metabolism in the body.

THE REGULATION OF INTERMEDIARY METABOLISM

Before considering the details of carbohydrate, lipid, and protein metabolism, it may help to outline briefly some of the mechanisms of regulation of metabolic processes. Factors that are involved include ATP and ADP, enzymes, coenzymes, and hormones. As discussed in the last chapter, energy flows from high-energy compounds such as 1,3-phosphoglyceric acid to ATP by means of the phosphorylation of ADP. The primary regulation of the utilization of the major body fuels is based on the effects of changes in ADP concentration on the rates of reactions that utilize ADP as a substrate. When more high-energy phosphate is used, more ADP is eventually produced, and a higher ADP concentration accelerates the reactions which utilize ADP and produce ATP.

Enzymes are involved in regulatory processes in intermediary metabolism. This regulation may occur at several levels, since the reaction rate of each enzyme system depends on the intracellular pH and the concentration of substrate, cofactors, and products, all of which are primary elements in the regulation of enzyme activity. Regulation of a specific metabolic sequence often depends on regulatory or **allosteric enzymes.** These enzymes are usually located at the beginning of a series of catabolic enzyme-catalyzed reactions; they catalyze the rate-limiting step of the series. As an example, a catabolic process that produces ATP as an end product may have its allosteric enzyme inhibited by ATP to control the rate of formation of this energy-rich compound. Some regulatory enzymes are activated or inhibited by the binding of a metabolite at a specific regulatory site. This could be achieved by competitive inhibition of a simple enzyme or by a more complex allosteric inhibition or activation of an enzyme containing multiple active sites. Another method of regulation of metabolic processes is by genetic control of the rate of enzyme synthesis. The mechanism of protein and enzyme synthesis will be discussed later in Chapter 12. The rate of any metabolic process depends on the concentration of each enzyme in the sequence of reactions; that concentration is the result of a balance between the rate of synthesis and the rate of degradation of the enzyme. Certain enzymes called **constitutive enzymes** are always present in fairly constant amounts in a particular cell. Other enzymes are synthesized in response to the presence of specific substrates, and are called **induced enzymes.** The genes responsible for the synthesis of these enzymes are usually repressed and only become active, or are derepressed, in the presence of the inducing agent, the specific substrate. This mechanism is discussed in more detail in Chapter 13.

The coenzymes NADP and NADPH are actively involved in the control of pathways in carbohydrate and lipid metabolism. The rate of the pentose phosphate pathway (p. 130) in carbohydrate metabolism is controlled by the availability of NADP. The initial reaction in the pathway uses glucose-6-phosphate dehydrogenase as its enzyme; experimental evidence indicates that the concentration of NADP limits the rate of the dehydrogenase. The pentose phosphate pathway converts NADP to NADPH. The reduced form of the coenzyme is used in many synthetic reactions in which it is converted back to NADP. The consumption of the NADPH in these reactions regulates the concentration of NADP, which in turn regulates the initial reaction in the pathway. These coenzymes are also involved in the synthesis of fatty acids from acetyl coenzyme A in lipid metabolism (p. 139); the ratio of the two forms exerts a control over this process.

Regulatory hormones produced by endocrine glands are chemical messengers that stimulate or inhibit specific metabolic processes in other tissues or cells. A familiar example of defective production of regulatory hormones is diabetes, where the pancreas is unable to secrete sufficient amounts of the hormone insulin to control the normal utilization of blood glucose (p. 121).

Catabolism of the major foodstuffs is always accompanied by conservation of some of the chemical energy in the form of ATP. This compound serves as a ready source of chemical energy to initiate catabolic and anabolic reactions in the cell.

IMPORTANT TERMS AND CONCEPTS

absorption	anabolism	digestion	Krebs cycle
acetyl coenzyme A	catabolism	Embden-Meyerhof pathway	metabolism

QUESTIONS

1. Outline the process of digestion and absorption of carbohydrates in the gastrointestinal tract.

2. Briefly describe the digestion and absorption of dietary fat.

3. Outline the process of protein digestion and absorption.

4. Define and compare the terms metabolism, catabolism, and anabolism.

5. Outline the general pathways of metabolism of the three major foodstuffs in the body.

6. What factors are involved in the regulation of intermediary metabolism in the cell? Explain.

Chapter 10

CARBOHYDRATE METABOLISM

The *objectives* of this chapter are to enable the student to:

1. Recognize the factors involved in the control of the normal blood sugar level.
2. Discuss the role of hormones in the control of the blood sugar level.
3. Describe the process of glycogenesis.
4. Discuss the role of cyclic-3′,5′-AMP and adenyl cyclase in the process of glycogenolysis.
5. Recognize the essential reactions in the Embden-Meyerhof pathway.
6. Outline the essential reactions in the Krebs cycle.
7. Account for the total number of moles of ATP formed in the Embden-Meyerhof and Krebs pathways.
8. Recognize the relationship between the phosphogluconate and the Embden-Meyerhof pathways.
9. Recognize the essential nature of chlorophyll in the light reaction of photosynthesis.
10. Outline the essential reactions in the dark reaction of photosynthesis.
11. Discuss the role of creatine phosphate in muscle contraction.

After the digestion of dietary polysaccharides and disaccharides to monosaccharides, these sugars are absorbed and carried by the portal circulation to the liver, mainly in the form of glucose, fructose, and galactose. The liver cells ordinarily convert all of the fructose and galactose into glucose, which is carried to the other tissues of the body in the bloodstream. The maintenance of the normal blood sugar level depends on many metabolic factors, which will be discussed in this chapter.

THE BLOOD SUGAR

After the monosaccharides are absorbed into the blood stream, they are carried by the portal circulation to the liver. Fructose and galactose are phosphorylated by liver enzymes and are either converted into glucose or follow similar metabolic pathways. The metabolism of carbohydrates, therefore, is essentially the metabolism of glucose.

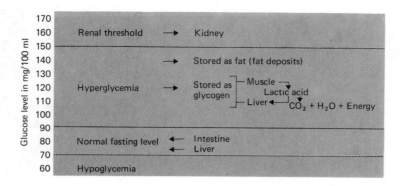

Figure 10-1 Factors involved in the regulation of the glucose level of the blood.

The concentration of glucose in the general circulation is normally 70 to 90 mg per 100 ml of blood. This is known as the **normal fasting level** of blood sugar. After a meal containing carbohydrates, the glucose content of the blood increases, causing a temporary condition of **hyperglycemia.** In cases of severe exercise or prolonged starvation, the blood sugar value may fall below the normal fasting level, resulting in the state of **hypoglycemia.** After an ordinary meal the glucose in the blood reaches hyperglycemic levels; this may be returned to normal by the following processes:

1. Storage
   (a) as glycogen
   (b) as fat
2. Oxidation to produce energy
3. Excretion by the kidneys

The operation of these factors in counteracting hyperglycemia is illustrated in Figure 10-1. The space between the vertical lines may be compared to a thermometer, with values expressed as milligrams of glucose per 100 ml of blood. During active absorption of carbohydrates from the intestine the blood sugar level rises, causing a temporary hyperglycemia. In an effort to bring the glucose concentration back to normal, the liver may remove glucose from the blood stream, converting it into glycogen for storage. The muscles will also take glucose from the circulation to convert it to glycogen or to oxidize it to produce energy. If the blood sugar level continues to rise, the glucose may be converted into fat and stored in the fat depots. These four processes usually control the hyperglycemia; but if large amounts of carbohydrates are eaten and the blood sugar level exceeds an average of 160 mg of glucose per 100 ml, the excess is excreted by the kidneys. The blood sugar level at which the kidney starts excreting glucose is known as the **renal threshold** and has a value from 150 to 170 mg per 100 ml.

In addition to the above factors, there are more specific reactions of the liver and the hormones to bring about regulation of the level of the glucose in the blood. The liver, for example, functions both in the removal of sugar from the blood and in the addition of sugar to the blood. During periods of hyperglycemia the liver stops pouring sugar into the blood stream and starts to store it as liver glycogen. During fasting the liver supplies glucose to the blood by breaking down its glycogen and by forming glucose from other food material such as amino acids or glycerol. The liver is assisted in this control process by several hormones.

HORMONES AND THE BLOOD SUGAR LEVEL

The properties and action of enzymes have already been discussed. In metabolism there are many related chemical reactions under the influence of enzymes. Another important group of regulating agents is the **hormones.** The hormones are formed mainly in the endocrine glands, which are also called ductless glands since their secretions diffuse directly into the blood stream. Enzyme action is more specific than hormone action, and the factors involved in the action of a hormone appear to be related to the action of other hormones. In a normal individual, major cellular processes depend on an endocrine or **hormone balance,** and a disturbance in this balance results in metabolic abnormalities. In the regulation of a body process, a hormone probably has control over several specific enzyme-catalyzed reactions.

Insulin

Although it was demonstrated as early as 1889 that removal of the pancreas of an animal would result in diabetes mellitus, it was not until 1922 that Banting and Best developed a method for obtaining active extracts of the pancreas. Within a short time insulin became available in sufficient quantities for the treatment of diabetes. It was first crystallized in 1926. More recently, as a result of the brilliant work of Sanger and his co-workers, a molecule consisting of two chains of amino acids with a molecular weight of 6000 has been described. The molecule synthesized by the pancreas is a long single-chain polypeptide called **proinsulin.** Within the pancreatic cells, prior to secretion of the hormone into the circulation, the proinsulin is converted to active insulin by forming disulfide crosslinks at the proper point in the chain and by the splitting off of an inactive peptide portion, called peptide C, by cellular enzymes. Since it is a protein, it is not effective when taken by mouth, because the proteolytic enzymes of the gastrointestinal tract cause its hydrolysis and destroy its activity. Insulin is usually injected subcutaneously when administered to a diabetic. In the adipose tissue cells, insulin causes an inhibition of adenyl cyclase (p. 125) and thus counteracts the effects of glucagon and epinephrine in the conversion of glycogen to glucose.

Insulin lowers the blood sugar level by increasing the conversion of glucose into liver and muscle glycogen, by regulating the proper oxidation of glucose by the tissues, and by preventing the breakdown of liver glycogen to yield glucose. In muscle and adipose tissue, insulin acts by increasing the rate of transport of glucose across membranes into the cells. Also, there is considerable evidence that in liver tissue insulin acts by controlling the phosphorylation of glucose to form glucose-6-phosphate, which is the first step in the formation of glycogen. In the absence of an adequate supply of insulin the transformation of extracellular glucose to intracellular glucose-6-phosphate is retarded.

TOPIC OF CURRENT INTEREST

DIABETES MELLITUS AND THE METABOLIC PROCESSES

Diabetes mellitus, or "sugar diabetes," has been recognized for many decades. Prior to the availability of insulin, beginning in the 1920s, patients with diabetes faced a progressive wasting disease, which culminated in death after a period of several months to a few years. Diabetes is a hereditary disease, characterized by insufficient production of insulin by the beta cells of the pancreas, resulting in an increase in blood sugar levels (hyperglycemia), glycosuria (excess glucose in the urine), and secondary alterations in fat and protein metabolism leading to tissue wasting, ketonuria (ketone bodies in urine), acidosis, and eventual coma and death, if uncontrolled. The central metabolic defect lies in the underutilization of glucose by muscle and adipose tissue as a consequence of an absolute or relative lack of insulin. An insufficiency of the capacity of the beta cells may be found after either pathological destruction or functional exhaustion of these cells. In addition to hereditary causes, an insulin deficiency may be caused by scarring of the pancreas after severe pancreatitis or by mumps.

Diabetes is usually a disease of middle life, with less than 10 per cent of known cases of diabetes classed as the juvenile type. The frequency of maturity-onset diabetes is highest in the sixth decade of life and is more common in women, especially after the menopause. Diabetes is more frequent among close relatives of known diabetics than among persons with no known diabetic relatives. There is a greater incidence of the disease among the siblings of diabetics than among their own children—when one identical twin has diabetes the other will also frequently develop the disease. In the majority of cases of spontaneous diabetes, heredity plays an etiological role. A diabetic trait predisposes persons to develop diabetes when either nonhormonal or hormonal factors impede the activity of otherwise normal amounts of insulin, and thereby interfere with the metabolism of carbohydrates and with intermediary metabolism in general. Obesity is an example of a nonhormonal factor that may trigger the onset of the disease, while excess growth hormone or glucocorticoids is an example of a hormonal factor.

Diabetes may exist in a latent form for several years before the disease becomes clinically apparent. It has been estimated that 15 to 20 per cent of the population are latent diabetics. The hyperglycemia in latent diabetes causes the pancreas to secrete excessive amounts of insulin, which produces a relative hypoglycemia 3 to 4 hours after a meal. This situation may continue for several years before the patient gradually begins to complain of generalized weakness and fatigue, weight loss, and the classic triad of symptoms of diabetes: polyphagia (excessive appetite), polyuria (excessive urination), and polydipsia (excessive thirst). This pattern is seen in the more common chronic form of the disease. There is also an acute-onset type, which may lead to a sudden disintegration of intermediary metabolism, with severe hyperglycemia, ketoacidosis, coma, and death. This acute type is far more common in juvenile diabetes than in the adult form.

Early diagnosis of the latent stage rather than active clinical stages is important, since the complications of diabetes may be minimized or perhaps prevented by early treatment. For this reason, considerable effort has been devoted to the development of tests for latent diabetes. When the proper dose of cortisone is administered just prior to a glucose tolerance test, individuals with latent diabetes exhibit a lower tolerance to the glucose than normal individuals. Other promising tests measure the concentration of insulin in the blood after administration of glucose (higher levels in latent diabetics) and a change in the adhesiveness of the blood platelets of latent diabetics after the addition of certain drugs to blood samples.

As diabetics are kept alive for many years by the administration of insulin and oral hypoglycemic agents, many long-term complications of the disease become apparent. These include changes in visual acuity with related retinopathy, lowered resistance to infections, pyorrhea, neuritis, glomerulonephritis, and atherosclerosis. Recently, there has been great concern that the present treatment of diabetes leaves much to be desired. According to Ellenberg, optimal control of diet, insulin, and exercise still does not prevent abnormal swings of blood sugar far above and below the range found in a normal person. Ellenberg states that diabetes is the main cause of blindness, and that more than one-half the number of heart attacks and three-fourths of strokes occurring are related to the disease. He claims that diabetes is the fifth leading cause of death by disease, and that when related complications are considered it may be the second leading cause. It is therefore apparent that this disease, which involves lipid and protein metabolism as well as carbohydrate metabolism, must be a subject of more intensive research.

Glucagon

Glucagon is a hormone that is produced by the α-cells of the pancreas. It is a polypeptide of known amino acid sequence with a molecular weight of about 3500. Glucagon causes a rise in the blood sugar level by increasing the activity of the enzyme liver phosphorylase, which is involved in the conversion of liver glycogen to free glucose. The activation of phosphorylase depends on the presence of the compound cyclic-3′,5′-adenosine monophosphate (AMP), whose formation from ATP by the enzyme adenyl cyclase is stimulated by the presence of glucagon.

Epinephrine

This hormone is produced by the central portion, or medulla, of the adrenal glands. Epinephrine is antagonistic to the action of insulin in that it causes glycogenolysis in the liver with the liberation of glucose. It stimulates the enzyme adenyl cyclase to produce cyclic-3′,5′-AMP from ATP and is also involved in the activation of phosphorylase. In addition to hyperglycemia, it also increases blood lactic acid by converting muscle glycogen to lactic acid. Continued secretion of epinephrine occurs under the influence of strong emotions such as fear or anger. This mechanism is often used as an emergency function to provide instant glucose for muscular work. The hyperglycemia that results often exceeds the renal threshold, and glucose is excreted in the urine.

Adrenal Cortical Hormones

Hormones such as **cortisone** and **cortisol** are produced by the outer layer or cortex of the adrenal gland. These hormones, especially those with an oxygen on position 11 (p. 84), have an effect on carbohydrate metabolism. In general they stimulate the production of glucose in the liver by increasing gluconeogenesis from amino acids. The cortical hormones are therefore antagonistic to insulin.

Anterior Pituitary Hormones

Of the many hormones secreted by the anterior lobe of the pituitary gland, the growth hormone, ACTH and the diabetogenic hormone affect the blood sugar level. The **growth hormone** causes the liver to increase its formation of glucose, but at the same time it stimulates the formation of insulin by the pancreas. Its action is complex and not completely understood. **ACTH,** the adrenocorticotropic hormone, stimulates the function of the hormones of the adrenal cortex and their action on the blood sugar level. The **diabetogenic hormone,** when injected into an animal, causes permanent diabetes and exhaustion of the islet tissue of the pancreas.

Although the overall control of the blood sugar level depends on the action of the liver and a balanced action of several hormones, it is readily apparent that insulin plays a major role in the normal process and is an important factor in the control of diabetes mellitus.

GLYCOGEN

As may be recalled from Chapter 5, glycogen is a polysaccharide with a branched structure composed of linear chains of glucose units joined by α-1,4 linkages and with α-1,6 linkages at the branch points. During absorption of the carbohydrates, the excess glucose is stored as glycogen in the liver. Normally this organ contains about 100g of glycogen, but it may store as much as 400g. The glycogen in the liver is readily converted into glucose and serves as a reservoir from which glucose may be drawn if the blood sugar level falls below normal. The formation of glycogen from glucose is called **glycogenesis,** whereas the conversion of glycogen to glucose is known as **glycogenolysis.** The muscles also store glucose as glycogen, but muscle glycogen is not as readily converted into glucose as is liver glycogen.

Glycogenesis

The process of glycogenesis is not just a simple conversion of glucose to glycogen. As we have learned previously, insulin is involved in the action of glucokinase in the phosphorylation of glucose to glucose-6-phosphate. The glucose-6-phosphate is then converted to glucose-1-phosphate with the aid of the enzyme phosphoglucomutase. The glucose-1-phosphate then reacts with uridine triphosphate (UTP) to form an active nucleotide, uridine diphosphate glucose (UDPG). In the presence of a branching enzyme and the enzyme UDPG-glycogen-transglucosylase, the activated glucose molecules of UDPG are joined in glucosidic linkages to form glycogen. These reactions may be represented as follows:

Uridine triphosphate (UTP)

Uridine diphosphate glucose (UDPG)

Glycogenolysis

The process of glycogenolysis liberates glucose into the blood stream to maintain the blood sugar level during fasting and to supply energy for muscular contraction. In the liver the reaction is initiated by the action of the enzyme phosphorylase, which splits the 1,4 glucosidic linkages in glycogen. The enzyme phosphorylase exists in two forms: an active form, **phosphorylase a,** and an inactive form, **phosphorylase b.** Phosphorylase b is converted to the active form of the enzyme by ATP in the presence of Mg^{+2} and phosphorylase b kinase, as shown:

$$2 \text{ Phosphorylase b} + 4 \text{ ATP} \xrightarrow[\text{cyclic-3',5'-AMP}]{\text{kinase, Mg}^{+2}} \text{Phosphorylase a} + 4 \text{ ADP}$$

The phosphorylase b kinase is activated by cyclic-3′,5′-AMP, a derivative of adenylic acid.

Cyclic-3′,5′-AMP

Cyclic-3′,5′-AMP is formed from ATP by the action of an enzyme called **adenyl cyclase.** This enzyme is activated by epinephrine and glucagon which therefore indirectly activate the phosphorylase responsible for the initiation of glycogenolysis. Other enzymes assist the breakdown to glucose-1-phosphate, which is subjected to the reversed action of phosphoglucomutase to yield glucose-6-phosphate. A specific enzyme in the liver, glucose-6-phosphatase, acts on glucose-6-phosphate to produce glucose. This enzyme is not present in muscle; therefore, muscle glycogen cannot serve as a source of blood glucose. These reactions may be illustrated as follows:

$$\text{Glycogen} \xrightarrow[\text{debranching enzyme}]{\text{phosphorylase}} \text{Glucose-1-phosphate}$$

$$\text{Glucose-1-phosphate} \xrightarrow{\text{phosphoglucomutase}} \text{Glucose-6-phosphate}$$

$$\text{Glucose-6-phosphate} \xrightarrow[\text{in liver}]{\text{glucose-6-phosphatase}} \text{Glucose}$$

OXIDATION OF CARBOHYDRATES

Glucose is ultimately oxidized in the body to form CO_2 and H_2O with the liberation of energy. **Glucose-6-phosphate** is a principal compound in the metabolism of glucose. As discussed earlier, it may be formed by the phosphorylation of glucose under the control of insulin. Once it is formed, it may be converted to glycogen or to free glucose, or it may be metabolized by several mechanisms or pathways. The two major pathways of glucose-6-phosphate metabolism are the **anaerobic,** or **Embden-Meyerhof, pathway** followed by the **aerobic,** or **Krebs, cycle.** The largest proportion of energy available from the oxidation of the glucose molecule is liberated from the Krebs cycle, but the Embden-Meyerhof pathway is essential in the formation of pyruvic acid used in the Krebs cycle.

GLYCOLYSIS

The ready availability of muscle preparations and their use in the development of physiology led to an early study of the biochemical changes associated with muscular contraction. It was observed that when a muscle contracts in an anaerobic medium, glycogen disappears and pyruvic and lactic acids are formed. In the presence of oxygen, or under aerobic conditions, the glycogen is re-formed, and the pyruvic and lactic acids disappear. Further studies demonstrated that one fifth of the lactic acid formed during glycolysis is oxidized to CO_2 and water, whereas the remaining four fifths is converted to glycogen.

Substances other than the carbohydrates in food and the lactic acid from muscular contraction may be converted into glycogen. These glycogenic compounds are formed by the process of **gluconeogenesis,** which is the conversion of non-carbohydrate precursors into glucose. Examples of these precursors are the glycogenic amino acids, the glycerol portion of fat, and any of the metabolic breakdown products of glucose, such as pyruvic acid, which may form glucose by reversible reactions in metabolism. The reactions discussed in the above section can be summarized in the **lactic acid cycle.**

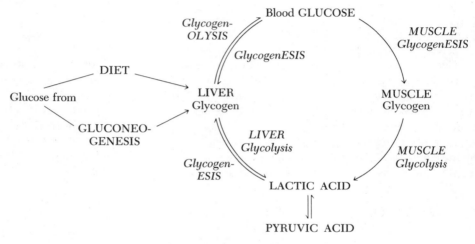

The Lactic Acid Cycle

Anaerobic, or Embden-Meyerhof, Pathway of Glycolysis

The chemical reactions in metabolic pathways like the Embden-Meyerhof pathway are detailed and complex, and may lead to confusion at first examination. An understanding of this and other metabolic pathways may be expedited by consideration of a preliminary outline of the essential reactions. Glucose from any source is converted to glucose-6-phosphate, which in turn is converted into fructose-6-phosphate, which is further phosphorylated to fructose-1,6-diphosphate. This set of reactions has converted the hexose glucose into a hexose diphosphate. The pathway diverges at this point with the splitting of fructose diphosphate into two triose monophosphates. One of these, glyceraldehyde-3-phosphate, is first transformed into 1,3-diphosphoglyceric acid, then successively into 3-phospho- and 2-phosphoglyceric acid. The last compound then forms phosphoenolpyruvic acid which is converted into pyruvic acid. The pyruvic acid is a key compound which may be reduced to lactic acid or further oxidized in the Krebs cycle.

The detailed chemical compounds and enzymes involved in the Embden-Meyerhof pathway are shown on the next page.

The requirement for and liberation of ATP in this anaerobic pathway is emphasized by the shaded areas. One mole of ATP is required for the phosphorylation of glucose and one more for the conversion of fructose-6-phosphate to fructose-1,6-diphosphate. The

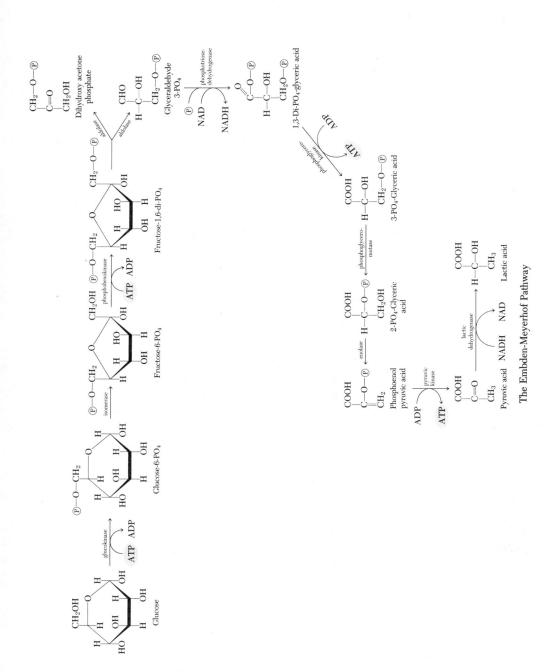

The Embden-Meyerhof Pathway

reaction of 1,3-diphosphoglyceric acid to form 3-phosphoglyceric acid liberates 1 mole of ATP per triose molecule or 2 moles of ATP per glucose molecule. The conversion of phosphoenolpyruvic acid to pyruvic acid also yields 2 moles of ATP per glucose molecule. For each mole of glucose broken down in the Embden-Meyerhof pathway, therefore, *2 moles of ATP are consumed and 4 moles are liberated, with a net gain of 2 moles of ATP.*

The two other common monosaccharides obtained from disaccharides in the diet, fructose and galactose, may also enter the scheme of glycolysis. Fructose may be phosphorylated at the 6 position by the enzyme hexokinase and then be converted to fructose-1,6-diphosphate. In the liver, fructose is phosphorylated at the 1 position by fructokinase. The fructose-1-phosphate is then converted into glyceraldehyde and dihydroxyacetone phosphate by a specific aldolase. The glyceraldehyde is further phosphorylated to glyceraldehyde-3-phosphate; these two triose phosphates enter the Embden-Meyerhof pathway. Galactose is phosphorylated at the 1 position by ATP and the enzyme galactokinase. In a series of reactions involving UTP, the galactose-1-phosphate is epimerized to glucose-1-phosphate, which is then converted to glucose-6-phosphate in the Embden-Meyerhof pathway. Regulation of glycolysis by the Embden-Meyerhof pathway occurs early in the chain of reactions. When the supply of ATP in the cell is above normal, the conversion of glycogen to glucose-1-phosphate is inhibited; when there is excessive concentration of glucose-6-phosphate, its formation from glucose is inhibited. These two processes control the formation of glucose-6-phosphate early in the pathway. A third control point is the inhibition of the conversion of fructose-6-phosphate to the 1,6-diphosphate in the presence of excess ATP. Conversely, when ADP is present in excess, the reaction proceeds through the regular pathway.

Aerobic, or Krebs, Cycle

The pyruvic acid formed in the Embden-Meyerhof pathway and the lactic acid from the lactic acid cycle or from the reduction of pyruvic acid are eventually oxidized with the formation of CO_2 and energy in the form of ATP. These reactions are carried out in the Krebs cycle, which may be outlined as follows.

Pyruvic acid from any source forms acetyl CoA, which transfers the acetyl group to oxaloacetic acid to make the tricarboxylic acid, citric acid. Citric acid is successively transformed into *cis*-aconitic acid, then to isocitric and to oxalosuccinic, all tricarboxylic acids. The hydration of *cis*-aconitic to isocitric acid as well as the later conversion of fumaric to malic acid involves the electrophilic addition of H^+. The oxalosuccinic acid then loses CO_2 to form α-ketoglutaric acid, which is converted to succinyl CoA and then to succinic acid and a series of dicarboxylic acids, including fumaric acid and malic acid, back to oxaloacetic acid, and the cycle is completed. The chemical structures, enzymes, and coenzymes of the Krebs cycle are shown on the next page.

The overall reaction for the conversion of pyruvic acid to carbon dioxide and water may be written as:

$$C_3H_4O_3 + \tfrac{5}{2}O_2 + 15ADP + 15P_i \rightarrow 3CO_2 + 2H_2O + 15ATP$$

The moles of ATP formed and CO_2 liberated in one turn of the Krebs cycle are shown in shaded areas. It may be recalled from a consideration of the electron transport mechanism in oxidative phosphorylation, page 109, that the oxidation of NADH or NADPH (through NADH) by way of the cytochrome system yields 3 moles of ATP per mole of NADH. Starting with FADH, the system yields 2 moles of ATP per mole of FADH.

When one molecule of glucose is completely oxidized it liberates 686.0 kilocalories. Each molecule of glucose subjected to the Embden-Meyerhof pathway liberates 8 moles

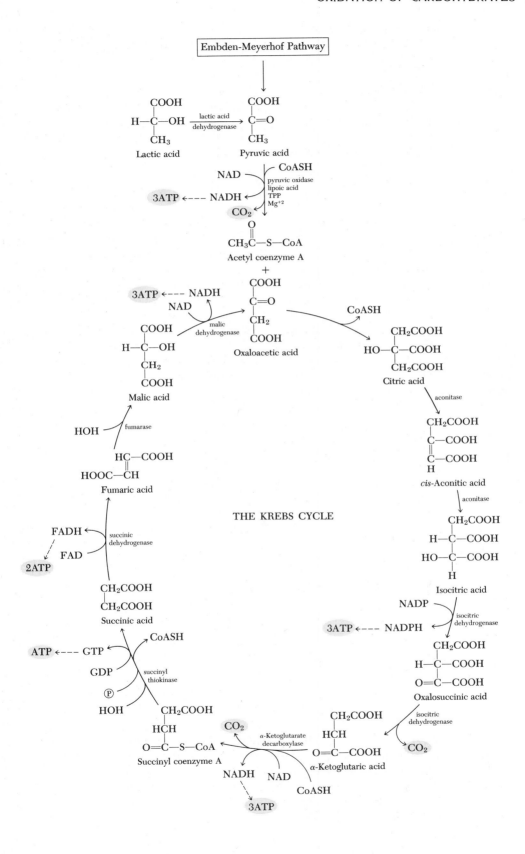

of ATP (6 moles from the NADH formed in the conversion of glyceraldehyde-3-phosphate to 1,3-diphosphoglyceric acid, and 2 moles net yield of ATP formed directly). Since each mole of glucose forms 2 moles of pyruvic acid, the Krebs cycle will yield 2×15 or 30 moles of ATP per molecule of glucose. *A total of 38 moles of ATP are therefore formed by the oxidation of a molecule of glucose in the Embden-Meyerhof and Krebs cycles.* Since each mole of ATP will yield approximately 8.0 kcal on hydrolysis, the 38 moles are equivalent to 304.0 kcal. This series of reactions is therefore capable of storing about 44 per cent of the available calories in the form of the high-energy compound ATP, to be used in muscular work and for other energy requirements.

ALTERNATE PATHWAYS OF CARBOHYDRATE OXIDATION

Pathways other than the Embden-Meyerhof and Krebs cycles are involved in the oxidation of carbohydrates. The most generally accepted alternate pathway is the **phosphogluconate pathway,** which is also called the **hexose monophosphate shunt,** or the **pentose phosphate pathway.** The key metabolic compound, glucose-6-phosphate, is oxidized to 6-phosphogluconolactone and then to 6-phosphogluconic acid. The acid is converted to a pentose, ribulose-5-phosphate, by the loss of CO_2, and the pentose is transformed into two other pentoses, xylulose-5-phosphate and ribose-5-phosphate. The two latter compounds are converted into a seven-carbon, sedoheptulose-7-phosphate, and the three-carbon glyceraldehyde-3-phosphate. These two compounds are then transformed into a four-carbon, erythrose-4-phosphate, and the hexose fructose-6-phosphate. The final reaction involves xylulose-5-phosphate and erythrose-4-phosphate, forming glyceraldehyde-3-phosphate and more fructose-6-phosphate.

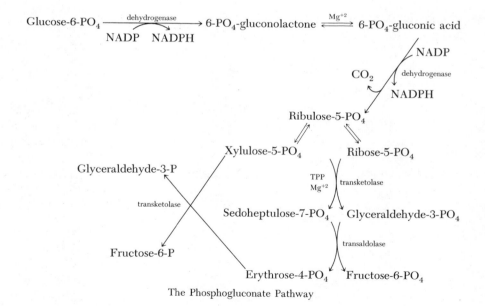

The Phosphogluconate Pathway

The primary purpose of this pathway in most cells is to generate reducing power in the cytoplasm in the form of NADPH. This coenzyme is particularly active in tissues that carry out the synthesis of fatty acids and cholesterol, such as liver, mammary gland, fat depots, and adrenal cortex. The pentoses that are formed, especially ribose, are used in the synthesis of nucleic acids, whereas the glyceraldehyde and fructose phosphates enter the Embden-Meyerhof pathway.

PHOTOSYNTHESIS

Carbohydrates are formed in the cells of plants from carbon dioxide and water. In the presence of sunlight and chlorophyll, the green pigment of leaves, these two compounds react to form pentoses, trioses, fructose, and more complex sugars. **Chlorophyll** is a protoporphyrin derivative containing magnesium that is located in the chloroplasts of green leaves.

Chlorophyll a

Originally the reaction between carbon dioxide and water to form carbohydrates was represented as follows:

$$CO_2 + H_2O \xrightarrow[\text{chlorophyll}]{\text{sunlight}} C_6H_{12}O_6 + 6\ O_2$$
$$\text{Simple sugar}$$

This process by which plants convert the energy of sunlight to form food material is called **photosynthesis.** Although photosynthesis is represented as a simple chemical reaction, it is more complex and includes several intermediates of the phosphogluconate and Embden-Meyerhof pathways.

The use of isotopes and radioactive tracers has greatly assisted the research workers in this field. For example, using isotopic oxygen, ^{18}O, as a tracer, it was shown that the oxygen liberated during photosynthesis is derived from the water molecules and not from carbon dioxide. When a green leaf is grown in an atmosphere of $^{14}CO_2$, the radioactive carbon appears very rapidly in a three-carbon atom and a five-carbon atom intermediate, and later in glucose and starch.

The Light Reaction

The reaction involving the conversion of light energy into chemical energy is called the **light reaction.** This transformation of energy occurs during photosynthetic phosphorylation and takes place in the chloroplasts of plants (Fig. 1–2). The essential reaction that occurs in **photophosphorylation** can be represented as follows:

$$ADP + P_i \xrightarrow{\text{light energy}} ATP$$

The photochemical process is initiated by the absorption of light by chlorophyll, which produces an excited-state molecule in which several electrons are raised from their normal energy level to a higher level in the double bond structure of chlorophyll. These excited electrons flow from chlorophyll to an iron-containing protein, **ferredoxin,** and bring about the reduction of NADP to form NADPH, which is used in the CO_2 fixation reactions of photosynthesis. Some of the excited electrons flow from ferredoxin through flavin pigments to a quinone structure called **plastoquinone,** then to **cytochrome pigments,** and then back to chlorophyll and their normal energy level. During this cycle some of the energy is given up by coupling in the reaction of ADP with P_i to form ATP. The ATP is also used in the CO_2 fixation reactions of photosynthesis. The electrons that are used in the formation of NADPH and ATP are replenished by a reaction in which the OH ions of water form molecular oxygen and donate electrons to chlorophyll through a cytochrome chain. The process of **photosynthetic phosphorylation** may be represented as follows:

The Process of Photophosphorylation in Photosynthesis

The Dark Reaction

The reactions of photosynthesis that are not dependent on light energy have been termed the **dark reaction.** This process involves the incorporation of carbon into carbohydrates, **carbon fixation,** and requires the energy from ATP and a quantity of NADPH formed in the light reaction. The dark reaction may be outlined as follows:

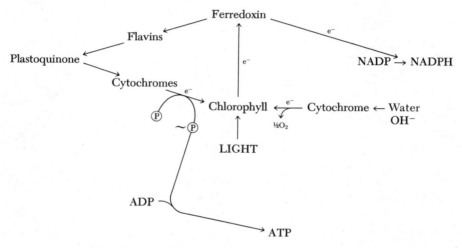

Ribulose-1,5-diphosphate plus carbon dioxide forms a complex that dissociates into two molecules of 3-phosphoglyceric acid, which is converted to 3-phosphoglyceraldehyde. This latter compound can be converted to glucose-6-phosphate through fructose-1,6-diphosphate and fructose-6-phosphate, and eventually to glucose and polysaccharides. The 3-phosphoglyceraldehyde may also be converted to dihydroxyacetone phosphate, which reacts with erythrose-4-phosphate to form sedoheptulose-7-phosphate, which reacts with glyceraldehyde-3-phosphate to form two molecules of pentose, ribose-5-phosphate and xylulose-5-phosphate. The pentoses can be converted into ribulose-1,5-diphosphate. The relationship between the hexose monophosphate shunt and the dark reaction is apparent from the common compounds, enzymes and coenzymes, especially NADPH. Some of the trioses and hexoses are also involved in the Embden-Meyerhof pathway, which illustrates the interlocking relationships that exist between the carbohydrate metabolic cycles.

MUSCLE CONTRACTION

Contraction of muscle fibers under anaerobic conditions leads to the formation of lactic acid and eventually to muscle fatigue. Muscle glycogen disappears during this process. In the presence of oxygen, the muscle regains its glycogen, loses lactic acid, and recovers its ability to contract. The high-energy compound **creatine phosphate** also changes form during contraction and subsequent recovery. The process of glycolysis supplies ATP, and the creatine phosphate and ATP join forces in muscular contraction.

Creatine phosphate   Creatine

It may help to recall that a direct linkage between nitrogen and phosphorus, as in creatine phosphate, denotes a high-energy compound. The reaction is readily reversible, and the muscle continues to contract as long as creatine phosphate is present.

During recovery, when more ATP is formed from glycolysis, the creatine phosphate is regenerated. The ATP is the direct source of energy for muscular work, and the function of creatine phosphate and glycolysis is to supply the ATP. Since there is only a small amount of ATP in the muscle at any instant, the supply of ATP needed for muscular work is obtained from creatine phosphate, the process of glycolysis, and the recovery of muscle glycogen. Thus muscular contraction is dependent on the cooperative action of several systems in carbohydrate metabolism.

IMPORTANT TERMS AND CONCEPTS

ATP formation
blood sugar level
chlorophyll
cyclic-3',5'-AMP
dark reaction
diabetes mellitus
Embden-Meyerhof pathway

epinephrine
glucose-6-phosphate
glycogenesis
glycogenolysis
glycolysis
insulin
Krebs cycle

lactic acid cycle
light reaction
phosphogluconate pathway

phosphorylase
photophosphorylation
photosynthesis

QUESTIONS

1. Discuss the factors involved in counteracting the normal hyperglycemia that occurs after a meal.

2. Discuss the role of insulin in the control of the normal blood sugar level.

3. What is latent diabetes, and why is it important to recognize this condition as early as possible?

4. List the hormones other than insulin involved in the control of the blood sugar level. Discuss the function of one of these hormones.

5. Describe the process of glycogenesis.

6. What is cyclic-3',5'-AMP? What role does it play in glycogenolysis?

7. How do the reactions of the lactic acid cycle explain the fate of the lactic acid formed by the process of glycolysis? Explain.

8. Outline the essential reactions in the Embden-Meyerhof pathway of glycolysis.

9. How many moles of ATP per glucose molecule are required to run the Embden-Meyerhof pathway, and how many moles are produced?

10. Outline the essential reactions in the aerobic Krebs cycle.

11. One turn of the Krebs cycle will yield how many moles of ATP? Why is this number so large compared to the Embden-Meyerhof pathway?

12. Outline the essential reactions in the phosphogluconate pathway.

13. What are the major products of the phosphogluconate pathway? How are they used in other pathways and cycles?

14. What is chlorophyll and what is its function in the process of photophosphorylation?

15. Explain how any three compounds formed in the CO_2 fixation scheme in photosynthesis could enter the reactions of the Embden-Meyerhof or Krebs cycle.

16. Outline the essential reactions in the light reaction of photosynthesis.

17. Outline the essential reactions in the dark reaction of photosynthesis.

Chapter 11

LIPID METABOLISM

The *objectives* of this chapter are to enable the student to:

1. Recognize the normal blood lipids and their role in metabolism and storage of fat.
2. Describe the reactions involved in the oxidation of fatty acids.
3. Account for the total number of moles of ATP formed in the oxidation of a fatty acid.
4. Outline the reactions involved in the synthesis of fatty acids.
5. Outline the reactions involved in the synthesis of triglycerides.
6. Describe the reactions that take place when ketone bodies are formed in the liver.
7. Discuss the synthesis of cholesterol and its relationship to atherosclerosis.
8. Describe the correlation between carbohydrate and lipid metabolism.

The major stores of energy-rich fuel in the cells are in the form of carbohydrates or fats. The energy for many of the body activities is derived from the metabolism of carbohydrates, as described in the last chapter. Triglycerides are also excellent sources of energy, since they have a caloric value more than twice that of carbohydrate or protein and are stored in fat depots in a nearly anhydrous form, compared to the hydrated molecules of glycogen in the cell. In a normal diet, up to 40 per cent of the caloric requirement is provided by dietary fat, whereas in fasting individuals or hibernating animals the fat stores supply almost all the energy required by the body.

Although the major energy source is fat, the metabolism of the lipids also involves phospholipids, glycolipids, and sterols. These latter substances are not stored in the fat depots but are essential constituents of tissues that play a role in fat transport and in many cellular metabolic reactions. These lipid derivatives also function as components of cell membranes, nerve tissue, membranes of subcellular particles such as microsomes and mitochondria, and chloroplasts in green leaves.

Unsaturated fatty acids such as **linoleic** and **linolenic** are essential components of cellular lipids that must be obtained in the diet, since they cannot be synthesized by the body. Cholesterol can be readily synthesized by the tissues and is currently a topic of considerable controversy, since there may be a relation between dietary cholesterol, blood cholesterol levels, and atherosclerosis.

BLOOD LIPIDS

The blood lipids to a certain extent parallel the behavior of the blood sugar. Their concentration in the blood increases after a meal and the level is returned to normal by processes of storage, oxidation, and excretion.

The lipids of the blood are constantly changing in concentration as lipids are added by absorption from the intestine, by synthesis, and by removal from the fat depots; they are removed by storage in the fat depots, oxidation to produce energy, synthesis to produce tissue components, and excretion into the intestine. The **normal fasting level** of blood lipids is usually measured in the plasma. Average values for young adults are as follows:

	mg/100 ml
Total lipids	510
Triglycerides	150
Phospholipids	200
Total cholesterol	160

The triglycerides, phospholipids, and cholesterol in the plasma are combined with protein as lipoprotein complexes. These **lipoproteins** are bound to the α- and β-globulin fractions of the plasma proteins and are transported in this form. A small amount of **nonesterified fatty acids** (**NEFA**) is always present in the blood and is bound to the albumin fraction of the plasma for transportation. These free fatty acids are thought to be the most active form of the lipids involved in metabolism. Their concentration is affected by the mobilization of fat from fat depots and by the action of several hormones.

FAT STORAGE

Fats may be removed from the blood stream by storage in the various fat depots. When fat is stored under the skin, it is usually called **adipose tissue.** However, considerable quantities of fat may be stored around such organs as the kidneys, heart, lungs, and spleen. This type of depot fat acts as a support for these organs and helps to protect them from injury. Recent studies employing the electron microscope reveal two major types of storage fat. One type is composed almost entirely of fat globules and has the characteristics of a storage depot. The second type contains many cells and a more extensive blood circulation, and is metabolically active, converting glycogen to fat and releasing fatty acids to other tissues as energy sources.

TOPIC OF CURRENT INTEREST

OBESITY: A DISEASE?

Obesity is a condition caused by the intake of more calories than the body requires over a period of time. In an adult, most of the food consumed is used to produce energy; food in excess of that necessary to fulfill the energy requirements of the body is efficiently converted to fat and stored in fat depots. In overweight individuals, this imbalance between food intake and energy expenditure has a tendency to become self-perpetuating. After years of overeating, the body weight gradually increases to the point where excess or even normal muscular activity is uncomfortable or unpleasant, thus contributing further to the imbalance of calorie intake versus calorie output. Another factor that may have a more general effect is that weight control may in some way be due to the appetite, which is abnormally increased in people who are gaining weight and decreased in those who are losing weight. There are obvious exceptions to these general statements, but they apply only to a very small percentage of obese individuals. There are a few people whose rate of metabolism decreases as they eat less food and who have great difficulty achieving the balance between the rate of muscular activity and food intake which maintains normal body weight. Obesity may rarely be due to a disorder of certain endocrine glands. An indication of the national awareness of the many problems of obesity is the proliferation of health spas, exercise salons, and publications of hundreds of diets designed to reduce the excess weight painlessly.

Unfortunately, the appetite level and propensity for over-eating may have been established by well-meaning parents during childhood. Parents who themselves are heavy eaters not only overload their dining tables with food but also encourage their children to eat heartily. By the time that the child is a young adult, appetite and eating habits have been firmly established. On the other hand, some people simply enjoy eating and fail to practice moderation. Other individuals reward themselves for minor achievements by consuming calorie-rich desserts.

Why do some biochemists, nutritionists, and physicians consider obesity a disease? In comparison with a person of the same age and normal body weight, the obese individual more frequently develops common metabolic diseases, especially in middle age. It has been clearly demonstrated that markedly overweight individuals have a higher incidence of maturity-onset diabetes, atherosclerosis, heart disease, hypertension, and circulatory problems leading to conditions such as phlebitis, along with greater severity and incidence of complications of rheumatoid arthritis and osteoarthritis. From a purely mechanical standpoint, the enlarged fat deposits crowd normal organs of the body such as the heart into abnormal positions, causing some impairment of function as well as a general impairment of circulation.

LIPOLYSIS

The role of hormones and cyclic-3′,5′-AMP in glycogenolysis (p. 125) has already been described. **Lipolysis,** the hydrolysis of triglycerides to free fatty acids, is also under the control of hormones and cyclic-3′,5′-AMP.

$$\text{Nonactivated lipase} + \text{ATP} \xrightarrow[\text{Cyclic-3′,5′-AMP}]{\text{Protein kinase}} \text{Activated lipase} + \text{ADP}$$

$$+$$

$$\text{Triglycerides}$$

$$\text{Free fatty acids}$$

$$+$$

$$\text{Glycerol}$$

Hormones such as epinephrine and glucagon stimulate the production of cyclic-3′,5′-AMP and therefore the process of lipolysis, while prostaglandins (p. 75) depress the levels of cyclic-3′,5′-AMP and decrease the rate of lipolysis.

OXIDATION OF FATTY ACIDS

Fatty acids that arise from the breakdown of any lipid, but especially from fats, are oxidized completely to form CO_2, water, and energy. The glycerol portion of fats is phosphorylated in the liver to form glycerophosphate, which is then oxidized to dihydroxyacetone phosphate.

$$\text{Glycerol} \xrightarrow[\text{ADP} \quad \text{ATP}]{\text{glycerokinase}} \text{Glycerophosphate} \xrightarrow[\text{NAD} \quad \text{NADH}]{\text{dehydrogenase}} \text{Dihydroxyacetone phosphate}$$

Both of these products can enter the Embden-Meyerhof pathway of carbohydrate metabolism.

The oxidation of fatty acids occurs in a series of reactions that require several enzymes and cofactors, with the production of acetyl coenzyme A. The acetyl CoA molecules then enter the Krebs cycle to form CO_2, H_2O, and energy. Early research by Knoop in 1904 established the fact that fatty acids were oxidized on the beta-carbon atom with the subsequent splitting off of two carbon fragments. In his **theory of beta-oxidation** he stated that acetic acid was split off in each stage of the process that reduced an 18-carbon fatty acid to a two-carbon acid.

In the past few years the detailed reactions, with their enzymes and cofactors, have been worked out, and Knoop's theory has been confirmed. Instead of acetic acid, the key compound in the reactions is acetyl CoA. Five reactions are involved in the conversion of a long chain fatty acid into a CoA derivative with two less carbon atoms and a molecule of acetyl CoA. These reactions are outlined in the scheme shown below.

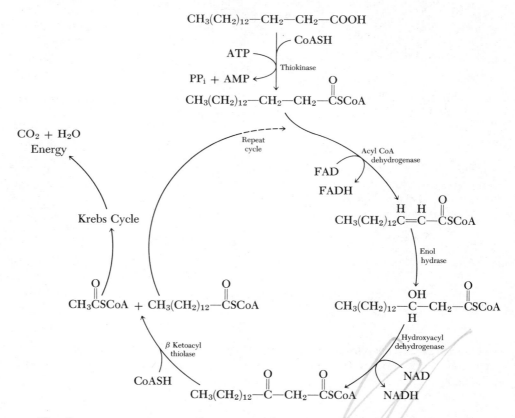

The first reaction initiates the series and involves the activation of a fatty acid molecule by conversion into a coenzyme A derivative. A dehydrogenase enzyme, with FAD as the coenzyme, desaturates the fatty acid; then a hydration is catalyzed by an enol hydrase. The hydroxyl group on the β-carbon atom is oxidized by a dehydrogenase with NAD as a coenzyme. The oxidized derivative plus coenzyme A is split into a fatty acid molecule with 2 less carbons, and acetyl CoA is formed. The acetyl CoA enters the Krebs cycle to form CO_2 and H_2O, plus energy. The new fatty-acid coenzyme A derivative does not have to be activated, but directly re-enters the cycle and again loses an acetyl CoA molecule. *Palmitic acid would require seven turns of the cycle to form 8 acetyl CoA moles.*

During the oxidation of palmitic acid, 7 FADH and 7 NADH moles would be formed. When these compounds enter the electron transport chain, they would form ATP as shown:

$$
\begin{array}{ll}
7\ \text{FADH} \rightarrow & 14\ \text{ATP} \\
7\ \text{NADH} \rightarrow & 21\ \text{ATP} \\
\hline
& 35\ \text{ATP} \\
- & \underline{1\ \text{ATP}}\ \text{used in first reaction} \\
& 34\ \text{Net ATP}
\end{array}
$$

In the seven turns of the cycle, 8 acetyl CoA moles are formed. As may be recalled from the oxidation of this compound in the Krebs cycle (p. 129), each mole of acetyl CoA will give rise to 12 moles of ATP. The acetyl CoA formed from the oxidation of palmitic acid will therefore account for the formation of $8 \times 12 = 96$ moles of ATP.

The sum of $34 + 96 = 130$ ATP for the complete oxidation of palmitic acid in the above scheme. The total combustion of palmitic acid yields 2338.0 kcal, and when compared to cellular oxidation (p. 130),

$$
\frac{130 \times 8.0\ \text{kcal} \times 100}{2338.0} = 48\%
$$

This represents a very efficient conservation of energy in the form of ATP molecules when palmitic acid is completely oxidized by the tissues. The previous discussion emphasizes the statement that food fat is an effective source of available energy. Also, a contributing factor to this efficient utilization is the fact that all the enzymes utilized in the β-oxidation scheme, the Krebs cycle, oxidative phosphorylation, and electron transport are found in the **mitochondria** of the cell.

SYNTHESIS OF FATTY ACIDS

The β-oxidation pathway in the mitochondria can be reversed to form fatty acid molecules, but this accounts for only a small percentage of the fatty acids synthesized in the tissues. The **cytoplasm** of the cell is the major site, and acetyl coenzyme A is the starting material, for the synthesis. Acetyl coenzyme A is carboxylated to form **malonyl coenzyme A** under the influence of acetyl CoA carboxylase in the presence of ATP and the vitamin biotin. An **enzyme-biotin complex** adds CO_2 with the help of ATP. Acetyl CoA then reacts with this complex to form malonyl CoA, as follows:

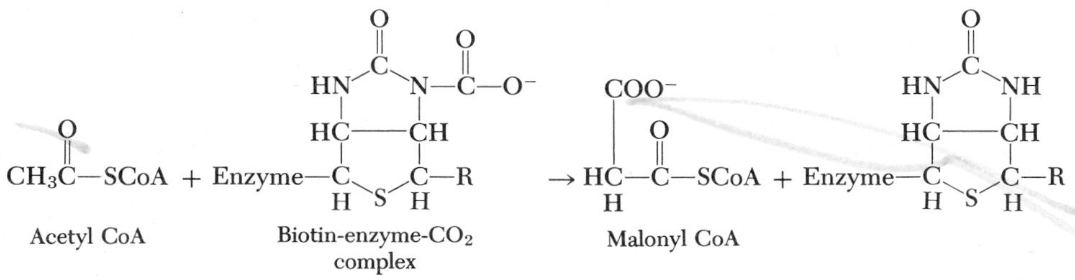

Malonyl CoA and acetyl CoA then form complexes with a multienzyme system called *fatty acid synthetase,* which includes an acyl carrier protein (ACP) that binds acyl intermediates during the formation of long-chain fatty acids. These two complexes then condense to form acetoacetyl-S-ACP, which is reduced to β-hydroxybutyryl-S-ACP with the assistance of NADPH, followed by the loss of a molecule of water to form an α,β-unsaturated-S-ACP. The unsaturated compound is reduced to butyryl-S-ACP, which combines with another molecule of malonyl-S-ACP to continue elongation of the chain.

SYNTHESIS OF TRIGLYCERIDES

Triglycerides are synthesized in the tissues from glycerol and fatty acids in activated forms. The active form of glycerol is L-glycerol-3-phosphate, which is formed from **dihydroxyacetone phosphate,** the product of the aldolase reaction in the Embden-Meyerhof pathway (p. 127). The glycerol phosphate reacts with two fatty acid CoA derivatives to form a **diglyceride,** which then reacts with another mole of fatty acid CoA to form a **triglyceride.** An outline of the process can be illustrated by the following scheme:

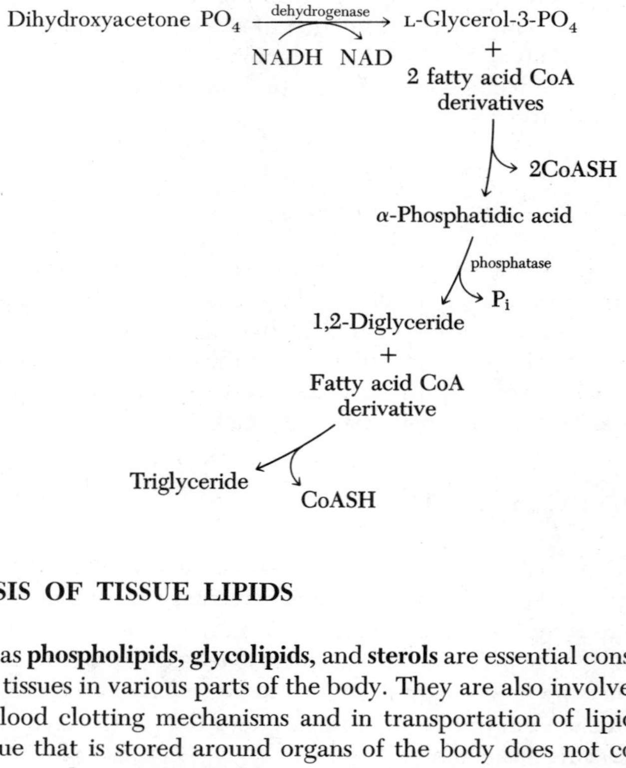

$$\text{Dihydroxyacetone PO}_4 \xrightarrow[\text{NADH} \quad \text{NAD}]{\text{dehydrogenase}} \text{L-Glycerol-3-PO}_4$$

+
2 fatty acid CoA
derivatives

$\searrow$ 2CoASH

α-Phosphatidic acid

phosphatase
$\searrow \text{P}_i$

1,2-Diglyceride
+
Fatty acid CoA
derivative

Triglyceride $\qquad$ CoASH

THE SYNTHESIS OF TISSUE LIPIDS

Lipids such as **phospholipids, glycolipids,** and **sterols** are essential constituents of cells, protoplasm, and tissues in various parts of the body. They are also involved in specialized functions, i.e., blood clotting mechanisms and in transportation of lipids in the blood. The adipose tissue that is stored around organs of the body does not contain the same proportion of saturated or unsaturated fatty acids as the food fat and therefore must also be synthesized. The most important organ in the body concerned with lipid synthesis is the liver. It is able to synthesize phospholipids and cholesterol and to modify all blood fats by lengthening or shortening, and saturating or unsaturating, the fatty acid chains.

Lecithin is used in transporting fats to the various tissues and may be involved in the oxidation of fats. Another essential phospholipid is **cephalin,** which is a vital factor in the clotting of blood. Special fats and oils in the body such as milk fat, various sterols, the natural oil of the scalp, and the wax of the ear are examples of lipids synthesized from the fats of the food.

FORMATION OF KETONE BODIES

The ketone, or acetone, bodies consist of **acetoacetic acid, β-hydroxybutyric acid, and acetone.** In a normal individual they are present in the blood in small amounts, averaging about 0.5 mg per 100 ml. Also, about 100 mg of ketone bodies is excreted per day in the urine. This low concentration in the blood and the small amount excreted in the urine are insignificant. But large amounts are present in the blood and urine during starvation and in the condition of diabetus mellitus. In general, any condition that results in a restriction of carbohydrate metabolism, with a subsequent increase in fat metabolism to supply the energy requirements of the body, will produce an increased formation of ketone bodies. This condition is called **ketosis.**

The precursor of the ketone bodies is acetoacetic acid which is formed in the liver from acetoacetyl CoA, a normal intermediate in the beta-oxidation of fatty acids. It may also be formed by the condensation of two molecules of acetyl CoA. Both methods of formation can be represented in the normal reversible reaction as follows:

$$2\ CH_3\overset{O}{\overset{\|}{C}}SCoA \underset{}{\overset{thiolase}{\rightleftharpoons}} CH_3\overset{O}{\overset{\|}{C}}CH_2\overset{O}{\overset{\|}{C}}SCoA + CoASH$$

Acetyl CoA   Acetoacetyl CoA

The liver contains a deacylase enzyme which readily converts acetoacetyl CoA to the free acid.

$$CH_3\overset{O}{\overset{\|}{C}}CH_2\overset{O}{\overset{\|}{C}}SCoA + H_2O \xrightarrow{deacylase} CH_3\overset{O}{\overset{\|}{C}}CH_2COOH + CoASH$$

Acetoacetyl CoA   Acetoacetic acid

The other ketone bodies are formed from acetoacetic acid: acetone by decarboxylation and β-hydroxy-butyric acid by the action of a specific enzyme, as shown in the accompanying diagram.

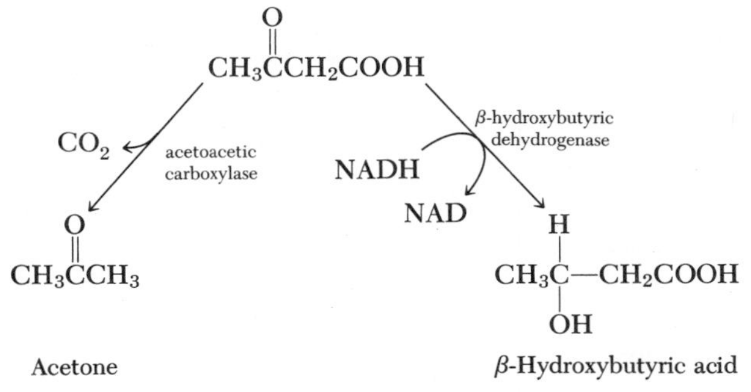

$$CH_3\overset{O}{\overset{\|}{C}}CH_2COOH$$

CO_2 ⤺ acetoacetic carboxylase

β-hydroxybutyric dehydrogenase

NADH

NAD

$$CH_3\overset{O}{\overset{\|}{C}}CH_3$$

Acetone

$$CH_3\overset{H}{\overset{|}{C}}-CH_2COOH$$
$$\underset{OH}{}$$

β-Hydroxybutyric acid

PHOSPHOLIPID METABOLISM

Knowledge concerning the metabolism of the phospholipids is incomplete, although they are known to serve many important functions in the body. Because their molecules are more strongly dissociated than any of the other lipids, they tend to be more soluble in water, to lower surface tension at oil-water interfaces, and to be involved in the electron transport system in the tissues. They would have a tendency to concentrate at cell membranes, and are probably involved in the transport mechanisms for carrying fatty acids and lipids across the intestinal barrier and from the liver and fat depots to other body tissues. Further evidence for their function in transporting lipids is found in their presence in the lipoproteins of the plasma. Phospholipids are essential components of the blood clotting mechanism, and sphingomyelin is one of the principal components of the myelin sheath of nerves.

Dietary phospholipids are probably broken into their constituents by enzymes in the gastrointestinal tract. The synthesis of most of the phospholipids has been established in recent years by the use of isotopes to tag precursors and intermediate compounds. For example, the synthesis of lecithin starts with the phosphorylation of choline by ATP; the phosphocholine then reacts with CTP to form the key compound **cytidine diphosphate choline,** which finally reacts with a 1,2-diglyceride. The phosphatidyl ethanolamines, or cephalins, are synthesized by similar reactions, and the sphingomyelins are synthesized by the reaction of N-acylsphingosine with cytidine diphosphate choline. Cytidine diphosphate choline (CDP-choline) has the structure:

Cytidine diphosphate choline (CDP-choline)

The synthesis of lecithin may be outlined as illustrated utilizing the 1,2-diglyceride as formed in triglyceride synthesis.

$$\text{1,2-Diglyceride} + \text{CDP-choline} \xrightarrow[\text{enzyme}]{\text{transferase}} \text{Lecithin} + \text{Cytidine monophosphate}$$

STEROL METABOLISM

The metabolism of sterols is mainly concerned with cholesterol and its derivatives. The **synthesis of cholesterol** and its relation to the other steroids of the body has been the subject of considerable research. Using either stable or radioactive isotopes, it has been shown that cholesterol can be synthesized from two-carbon compounds such as acetyl CoA. It can also be synthesized from acetoacetyl CoA and other intermediates. Two of the intermediate cholesterol precursors are β-hydroxy-β-methylglutaryl-SCoA and mevalonic acid, as shown on the following page. Although the synthesis of cholesterol occurs in many tissues in the body, the liver is the main site of cholesterol formation.

Cholesterol is a key compound in the synthesis of essential steroids such as bile acids, sex hormones, adrenal cortical hormones, and vitamin D. Not only is cholesterol converted to bile acids by the liver, but it is also excreted as such in the bile. In addition, the cholesterol in the bile can give rise to gallstones by accumulating on insoluble

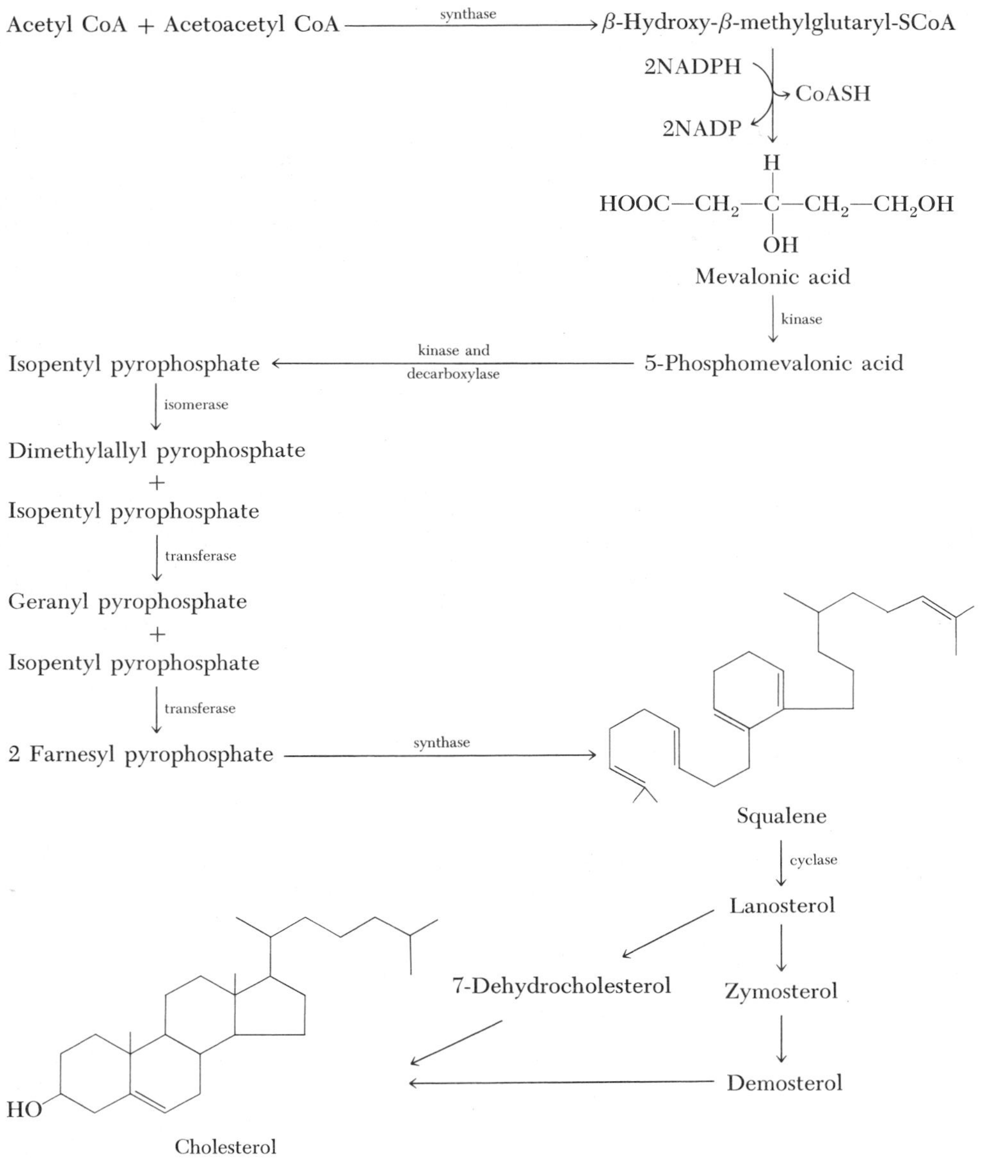

Acetyl CoA + Acetoacetyl CoA ——————synthase——————→ β-Hydroxy-β-methylglutaryl-SCoA

2NADPH ⟶ CoASH

2NADP

$$HOOC-CH_2-\underset{\underset{OH}{|}}{\overset{\overset{H}{|}}{C}}-CH_2-CH_2OH$$

Mevalonic acid

↓ kinase

Isopentyl pyrophosphate ←——kinase and decarboxylase—— 5-Phosphomevalonic acid

↓ isomerase

Dimethylallyl pyrophosphate
+
Isopentyl pyrophosphate

↓ transferase

Geranyl pyrophosphate
+
Isopentyl pyrophosphate

↓ transferase

2 Farnesyl pyrophosphate ——synthase——→

Squalene

↓ cyclase

Lanosterol

↓

7-Dehydrocholesterol   Zymosterol

↓

Demosterol

HO

Cholesterol

objects or particles. The concentration of cholesterol in the blood is apparently dependent on the dietary intake of sterols and neutral fats, and the synthesis of cholesterol by the liver. The normal level in the blood gradually increases with age and ranges from 150 to 200 mg/100 ml. Blood cholesterol levels are often determined in patients to assess their cholesterol status. Many methods have been devised for this determination, and many of them are modifications of the Liebermann-Burchard reaction described in Chapter 6. If the cholesterol level in the blood is maintained at an abnormally high concentration, such as 200 to 300 mg/100 ml, deposition of cholesterol plaques may occur in the aorta and lesser arteries. This condition, known as **atherosclerosis** or **arteriosclerosis,** is seen in older persons and often results in circulatory or heart failure. Considerable research effort is being directed at this problem in an attempt to reduce the cholesterol level in the blood of these patients and thus alleviate the symptoms of the disease.

CORRELATION OF CARBOHYDRATE AND FAT METABOLISM

From a nutritional standpoint it has long been apparent that carbohydrate can be converted into fat in the body. When glucose tagged with ^{14}C was fed to animals, the fatty acids of liver and other tissue fat were found to be labeled with ^{14}C. The conversion of fat to carbohydrate has long been open to question. The glycerol portion of fat is closely related to the three-carbon intermediates of carbohydrate metabolism, but it has been more difficult to demonstrate a direct relation between fatty acids and glucose. Since the role of acetyl CoA has been established in both carbohydrate and fat metabolism, it is apparent that the acetyl CoA from fatty acid oxidation can enter the Krebs cycle

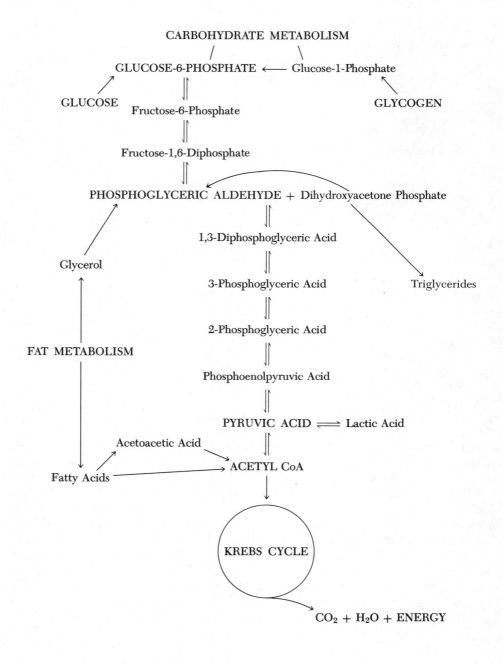

in the same fashion as this compound formed from pyruvic acid. More recently it has been shown that ^{14}C labeled fatty acids are converted to ^{14}C labeled glucose in a diabetic animal. The correlation between carbohydrate and fat metabolism in the body may be represented in the scheme shown on the opposite page.

IMPORTANT TERMS AND CONCEPTS

acetyl CoA
adipose tissue
ATP formation
blood lipids
cholesterol synthesis

fatty acid oxidation
fatty acid synthesis
ketone bodies
obesity
triglyceride synthesis

QUESTIONS

1. List the important lipids in a normal individual's blood and their approximate concentration.

2. Briefly discuss the disadvantages of obesity.

3. Outline the essential reactions in the scheme for oxidation of fatty acids.

4. How do you account for the large number of moles of ATP formed in the oxidation of fatty acids?

5. Outline the reactions involved in the synthesis of fatty acids.

6. Describe the process of synthesis of triglycerides.

7. Discuss two reactions that may take place in the liver for the formation of acetoacetic acid.

8. What is ketosis? Explain the cause of this condition in the body.

9. Show in outline form the compounds involved in the synthesis of cholesterol.

10. Why is atherosclerosis receiving so much attention in our society?

Chapter 12

PROTEIN METABOLISM

The *objectives* of this chapter are to enable the student to:

1. Recognize the relationship between the amino acid pool and the dynamic state of tissue proteins.
2. Discuss the role of essential amino acids in protein synthesis during growth.
3. Describe the activation of amino acids and their combination with t-RNA prior to protein synthesis.
4. Discuss the role of ribosomes in protein synthesis.
5. Illustrate the processes of transcription and translation in protein synthesis.
6. Explain the relationship between codons and anticodons in protein synthesis.
7. Recognize the difference between deamination and transamination.
8. Outline the essential reactions in the urea cycle.
9. Recognize the reactions involved in purine and pyrimidine metabolism.
10. Describe the correlation between carbohydrate, lipid, and protein metabolism.

In the metabolism of proteins in the living cell, amino acids may serve as precursors of proteins or they may be oxidized as a source of energy. The synthesis of new proteins for growth and development, or anabolism, involves the building of different amino acids into the proper sequences and spatial arrangements to produce specific protein molecules. Tissue proteins of various species of animals, plants, and microorganisms all have specific structures and compositions. Protein enzymes and hormones, plasma proteins, and the protein of hemoglobin and of various nucleoproteins represent other types of proteins. As has been discussed in the preceding chapters, a fasting individual and a patient with diabetes mellitus utilize their fat depots as a source of energy. Tissue proteins are also utilized as a source of fuel in these situations. The process of catabolism of proteins to produce energy involves many general metabolic reactions and many that are specific for the metabolism of one of the 22 different amino acids.

AMINO ACID POOL

In contrast to carbohydrate and fat metabolism, there are no storage depots for proteins or amino acids. The increased concentration of amino acids that occurs from the process of absorption, synthesis, or catabolism represents a temporary pool of amino acids

146

which may be used for metabolic purposes. This pool of amino acids is available to all tissues and may be synthesized into new tissue proteins, blood proteins, hormones, enzymes, or nonprotein nitrogenous substances such as creatine and glutathione. The relationship that exists between the **amino acid pool** and protein metabolism in general may be represented as follows:

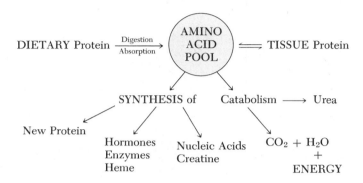

THE DYNAMIC STATE OF BODY PROTEIN

Until the late 1930s it was believed that the body proteins of the adult human were stable molecules and that the majority of the amino acids from the diet were catabolized to produce energy. A small proportion was thought to be used for maintenance and repair of the existing tissue proteins. When isotopes became available, Schoenheimer and his associates demonstrated that tissue proteins exist in a *dynamic state of equilibrium*. When the nitrogen of an amino acid was labeled with ^{15}N and incorporated in the diet of an animal, about 50 per cent of the ^{15}N was found in the tissues of the animal, and a greater percentage was found in the nitrogen of amino acids other than that specifically fed. This indicated that the amino acids of tissue proteins were constantly changing places with those in the amino acid pool, and that the body proteins were extremely labile molecules.

More recent research using isotopically labeled amino acids has shown that tissue proteins vary considerably in their rate of turnover of amino acid molecules. The **turnover rate** represents the amount of protein synthesized or degraded per unit time, and the **turnover time** is usually expressed as the half-life of a protein in the tissues. Liver and plasma proteins have turnover times (a half-life) of 2 to 10 days, in contrast to 180 days for muscle protein and 1000 days for some collagen proteins. Muscle and connective tissue proteins appear to have a very prolonged turnover compared to liver and plasma proteins, which are rapidly synthesized from the amino acids in the amino acid pool. The concept of the dynamic state of body protein requires modification in view of the individuality of specific proteins.

THE SYNTHESIS OF PROTEIN

The process of synthesis of protein is always occurring in the body, especially during growth and in those tissues with a rapid turnover rate.

ESSENTIAL AMINO ACIDS AND PROTEIN SYNTHESIS

The synthesis of protein is always occurring in the body, especially in those tissues with a rapid turnover rate. A growing child or animal is continually building new tissue and therefore makes the greatest demand on the amino acid pool. The individual amino acids required for protein synthesis are apparently sorted out by the body and used to construct specific protein molecules. Although the tissues, particularly the liver, are able to synthesize some amino acids, others must be present in the diet to assure a complete and proper assortment for synthetic purposes.

The amino acids that cannot be synthesized by the body and must therefore be supplied by dietary protein are called **essential amino acids.** If an essential amino acid is lacking in the diet, the body is unable to synthesize tissue protein. If this condition occurs for any length of time, a negative nitrogen balance will exist, and there will be weight loss, lowered serum protein level, and marked edema. Extensive feeding experiments on laboratory rats have established the following amino acids as essential for growth:

Histidine	Isoleucine
Methionine	Leucine
Arginine	Lysine
Tryptophan	Valine
Threonine	Phenylalanine

From the studies of Rose on the amino acid requirements of man, it was proposed that all of the above ten except histidine and possibly arginine are essential to maintain nitrogen balance. An individual is in **nitrogen balance** when the nitrogen excreted equals the nitrogen intake in a given period of time. A growing child or a patient recovering from a prolonged illness is in **positive nitrogen balance.** Starvation, a wasting disease, or a diet lacking sufficient amounts of essential amino acids can result in a **negative nitrogen balance.**

Many common dietary proteins are deficient in one or more of these essential amino acids. Gelatin, for example, lacks tryptophan and is therefore an **incomplete protein.** If gelatin were the sole source of protein in the diet, a growing child could eat large quantities of this protein every day without building new body tissues. Zein and gliadin, the prolamines of corn and wheat respectively, are deficient in lysine, and zein is also low in tryptophan. The supplementing of these proteins with small amounts of the deficient amino acid (lysine or tryptophan) will result in adequate growth. Although an incomplete protein will not support growth when it is the only protein in the diet, we seldom confine ourselves to the consumption of a single protein. In an ordinary mixed diet the essential amino acids are best supplied by protein of animal origin, such as meat, eggs, milk, cheese, and fish.

In other parts of the world, particularly in Africa, Asia, and South America, the diet is often deficient in high quality protein. Subnormal growth in children and prevalence of disease and early death are common in these populations. Small children develop **kwashiorkor** when deprived of adequate protein, and their bloated bellies and wrinkled skin are often seen in photographs pleading for food for hungry nations. Many of these children die before they can be adequately nourished. Agricultural research aimed at the production of high-yield wheat, high-lysine corn, and edible protein from fish meal and seaweed is being conducted to increase the world supply and availability of protein food.

Mechanism of Protein Synthesis

A very active field of research at the present time is the study of the mechanism of protein synthesis, the sequence of amino acids in the protein being synthesized, and the nature of the genetic code responsible for this sequence. Protein synthesis is initiated by the activation of amino acids. This process occurs through the combination of the amino acid with ATP and an enzyme specific for the amino acid, with the splitting off of two molecules of phosphoric acid.

$$R\text{—}CH\text{—}COOH \xrightarrow[\text{ATP} \quad \text{PP}_i]{\text{aminoacyl synthetase}} Enz\text{-}Adenine\text{-}Ribose\text{—}O\overset{\overset{O}{\|}}{\text{—}P}\text{—}O\overset{\overset{O}{\|}}{\text{—}C}\text{—}CH\text{—}R$$

$$\underset{\text{Amino acid (AA)}}{\underset{NH_2}{|}} \qquad\qquad\qquad \underset{\substack{\text{Amino acyl AMP enzyme complex}\\ \text{(E-AMP-AA)}}}{\underset{NH_2}{|}}$$

The second step of the process involves the transfer of the activated amino acid to a specific **transfer RNA molecule.** The t-RNA molecules are small (mol. wt. 30,000) nucleic acid molecules with a terminal grouping of cytidylic-cytidylic-adenylic acid, C-C-A, which serves as the binding site for the amino acid. Each molecule of t-RNA must have two distinct areas of nucleotide sequence: one to contain the anticodon, which binds specific codons on m-RNA, the other to form bonds with the enzyme that catalyzes the attachment of the activated amino acid to its specific t-RNA at the C-C-A terminus. A three-dimensional representation of a transfer RNA indicating active areas of the molecule is shown in Figure 12–1. The transfer of the activated amino acid is under the influence of the same enzyme that produced the activation and cytidine triphosphate (CTP).

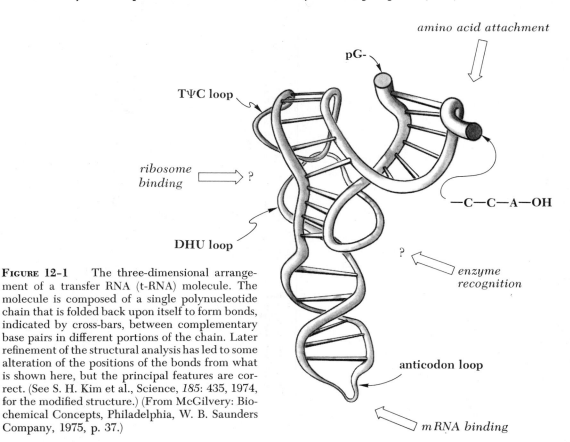

Figure 12–1   The three-dimensional arrangement of a transfer RNA (t-RNA) molecule. The molecule is composed of a single polynucleotide chain that is folded back upon itself to form bonds, indicated by cross-bars, between complementary base pairs in different portions of the chain. Later refinement of the structural analysis has led to some alteration of the positions of the bonds from what is shown here, but the principal features are correct. (See S. H. Kim et al., Science, *185*: 435, 1974, for the modified structure.) (From McGilvery: Biochemical Concepts, Philadelphia, W. B. Saunders Company, 1975, p. 37.)

$$\text{E-AMP-AA} + \text{t-RNA} \longrightarrow \text{t-RNA-AA} + \text{AMP} + \text{E}$$

| (activated amino acid) | transfer RNA | transfer RNA-amino acid complex |

This enzyme must have two binding sites: one to associate with the enzyme recognition site of the t-RNA (mentioned above), the other to bind a specific amino acid from the mixture in solution in the cell and catalyze its attachment to the t-RNA.

The third step involves the transfer of t-RNA-bound amino acids to the ribosomes of the cellular cytoplasm. **Ribosomes** are nucleoprotein particles composed of approximately 60 per cent basic proteins and 40 per cent ribosomal RNA (r-RNA). There are two subunits, a 30 S and a 50 S ribosome, that are joined to make a complete 70 S ribosome in the bacterial cell *E. coli*. In mammalian cells, the two subunits are a 40 S and a 60 S ribosome. In the cytoplasm, ribosomes are joined by strands of messenger RNA to form polysomes, which consist of 4 to 6 ribosomes. Both the t-RNA and m-RNA molecules are bound by the ribosomes to achieve a synthesis of polypeptide from the activated amino acids carried by the t-RNA.

The active template that controls protein synthesis on the ribosomes is called **messenger RNA** (m-RNA). It is synthesized by RNA polymerase under the direction of DNA in the nucleus, and it carries information to the ribosomes to direct the sequence of alignment of t-RNA-bound amino acids. The formation of m-RNA requires, in addition to the enzyme RNA polymerase, a mixture of the triphosphates of the nucleosides to be incorporated in the m-RNA molecule (see Fig. 12–2). A representative ribosome with its characteristic features and specific sites is also shown in Figure 12–2. This first process of overall

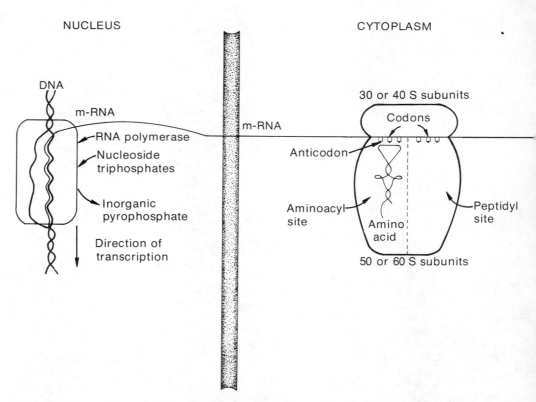

Figure 12-2   A representation of the process of transcription involving DNA, m-RNA, RNA polymerase and a ribosome with its characteristic features and active sites.

TABLE 12-1 CODONS FOR SPECIFIC
AMINO ACIDS

AMINO ACIDS	CODONS FOR m-RNA
Alanine	GCA, GCC, GCG, GCU
Arginine	AGA, AGG, CGA, CGG, CGC, CGU
Asparagine	AAC, AAU
Aspartic acid	GAC, GAU
Cysteine	UGC, UGU
Glutamic acid	GAA, GAG
Glutamine	CAG, CAA
Glycine	GGA, GGC, GGG, GGU
Histidine	CAC, CAU
Isoleucine	AUA, AUC, AUU
Leucine	CUA, CUC, CUG, CUU, UUA, UUG
Lysine	AAA, AAG
Methionine	AUG—Chain initiation
Phenylalanine	UUU, UUC
Proline	CCA, CCC, CCG, CCU
Serine	AGC, AGU, UCA, UCG, UCC, UCU
Threonine	ACA, ACG, ACC, ACU
Tryptophan	UGG
Tyrosine	UAC, UAU
Valine	GUA, GUG, GUC, GUU
Chain termination	UAA, UAG, UGA

information transfer from DNA to m-RNA is called **transcription** (transcribing the message). The specific programming of the amino acids on the polysomes or m-RNA molecules to synthesize a protein containing a definite sequence of amino acids is called **translation** (translating the code).

The attachment of the activated amino acids bound to t-RNA to a specific area or site on the m-RNA is governed by the genetic code. There is a specific site on m-RNA consisting of three consecutive nucleotide residues or bases that binds a particular amino acid; this site is called a **codon.** The t-RNA for that particular amino acid (see Figure 12–1) has a complementary triplet of bases called an **anticodon** that binds the t-RNA to the site on the m-RNA on the ribosome. Since there are four different nucleotide residues or bases in m-RNA, and a sequence of three bases is involved in coding for each amino acid, a total of 4^3 or 64 different combinations of the three bases is available for coding. The genetic code is **nonoverlapping** in that it requires the action of a specific group of three bases on the m-RNA chain, and it is also said to be **degenerate** in that more than one codon may be employed by m-RNA to insert a specific amino acid into the peptide chain. The codons currently proposed for specific amino acids are shown in Table 12–1. Note that there is a specific codon for the initiation of a protein chain and three chain termination codons.

Each ribosome has two binding sites on its 50 S or 60 S subunit. One, the aminoacyl site, binds a specific t-RNA, carrying its attached amino acid; the other binds the t-RNA attached to the growing peptide chain and is called the peptidyl site. A sequence in the synthesis of a polypeptide chain by a polyribosome attached to a strand of m-RNA is shown in Figure 12–3. The first ribosome in the series has a small polypeptide attached to t-RNA$_{ser}$ on the peptidyl site and a t-RNA$_{ala}$ with its amino acid is shown being attached to the aminoacyl site. With the assistance of the proper enzymes and cofactors, the peptide chain is transferred to the alanine on the t-RNA$_{ala}$ and the growing chain moves to the peptidyl site on the ribosome. A new t-RNA for glycine then attaches to the aminoacyl site and the process of elongation of the polypeptide chain continues.

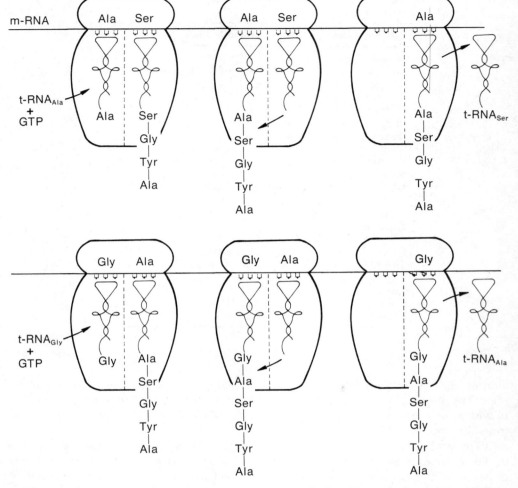

Figure 12–3   A sequence in the synthesis of a polypeptide chain by a polyribosome attached to a strand of m-RNA.

METABOLIC REACTIONS OF AMINO ACIDS

The amino acids in the metabolic pool that are not immediately used for synthesis can undergo several metabolic reactions. They may follow the path of catabolism through deamination, urea formation, and energy production, or they may assist in the synthesis of new amino acids by the process of reamination and transamination. **Deamination, reamination, transamination,** and **urea formation** are processes common to all amino acids and are therefore very important to protein metabolism.

Deamination

A general reaction of catabolism is the splitting off of the amino group of an amino acid, with the formation of ammonia and a keto acid. This process is called **oxidative deamination** and is catalyzed by enzymes found in liver and kidney tissue called **amino acid oxidases.** These enzymes are generally flavoprotein enzymes containing either flavin adenine dinucleotide, FAD, or flavin mononucleotide, FMN. The enzyme dehydrogenates the amino acid to form an imino acid, which is hydrolyzed to a keto acid and ammonia. The process may be illustrated with a type formula for an amino acid.

$$R-\underset{\underset{\text{Amino acid}}{NH_2}}{\overset{}{\underset{|}{CH}}}-COOH \xrightarrow[\text{oxidase}]{\text{amino acid}} \underset{FAD \qquad FADH}{\longrightarrow} R-\underset{\underset{\text{Imino acid}}{NH}}{\overset{\|}{C}}-COOH \xrightarrow[H_2O]{\text{hydrolysis}} R-\underset{\underset{\text{Keto acid}}{O}}{\overset{\|}{C}}-COOH + NH_3$$

The fate of the keto acid depends on the amino acid from which it is derived. In general the catabolism of each amino acid must be studied separately. Glycine, for example, is the simplest amino acid, yet it can be transformed metabolically to formate, acetate, ethanolamine, serine, aspartic acid, fatty acids, ribose, purines, pyrimidines, and protoporphyrin. This amino acid may therefore play a role in carbohydrate, lipid, protein, nucleic acid, and hemoglobin metabolism, and it admirably illustrates the interrelationships that exist among the different types of metabolism in the body. Other amino acids undergo complex metabolic reactions that are beyond the scope of this book. In general, amino acids are classed as **glucogenic** when their carbon atoms form pyruvate or intermediates in the Krebs cycle which can be converted to glucose. Alanine, serine, threonine, aspartic and glutamic acids, asparagine, and glutamine are examples of glucogenic amino acids. Other amino acids such as valine, leucine, and isoleucine are said to be **ketogenic**, since they readily form acetyl groups and acetoacetate which follow the metabolic pathway of lipids and form ketone bodies.

Transamination

The process of deamination results in the formation of many keto acids that are capable of accepting an amino group to form a new amino acid. That this process of **reamination** occurs was readily apparent from the isotope-labeling experiments of Schoenheimer. He observed a ready exchange of amino groups of dietary amino acids and tissue amino acids. A major mechanism for the conversion of keto acids to amino acids in the body is known as **transamination**. Transaminations are catalyzed by enzymes known as **transaminases** or **aminotransferases**. Since most of these enzymes require α-ketoglutaric acid as the amino group receptor, they are specific for the substrate pair of α-ketoglutaric and glutamic acids. The important transaminase found in many tissues is aspartate transaminase or aspartate aminotransferase, which catalyzes the following reaction:

$$
\underset{\text{Aspartic acid}}{\begin{array}{c}COOH\\|\\CH_2\\|\\HCNH_2\\|\\COOH\end{array}} + \underset{\alpha\text{-Ketoglutaric acid}}{\begin{array}{c}COOH\\|\\CH_2\\|\\CH_2\\|\\C=O\\|\\COOH\end{array}} \xrightarrow[\substack{\text{pyridoxal}\\\text{phosphate}}]{\substack{\text{aspartate}\\\text{transaminase}}} \underset{\text{Oxalacetic acid}}{\begin{array}{c}COOH\\|\\CH_2\\|\\C=O\\|\\COOH\end{array}} + \underset{\text{Glutamic acid}}{\begin{array}{c}COOH\\|\\CH_2\\|\\CH_2\\|\\HCNH_2\\|\\COOH\end{array}}
$$

Other amino acids such as alanine, leucine, or tyrosine may replace aspartic acid in the above reaction. Alanine, for example, with the enzyme alanine transaminase, would react with α-ketoglutaric acid to form pyruvic acid and glutamic acid. The coenzyme **pyridoxal phosphate** is required in the reaction and **pyridoxamine phosphate** is formed.

Pyridoxal phosphate   Pyridoxamine phosphate

$$R_2-\underset{\underset{O}{\|}}{C}-COOH \;+\; CH_2NH_2 \quad \dashrightarrow \quad R_2-\underset{\underset{N}{\|}}{C}-COOH \;/\; CH \quad \longrightarrow \quad R_2-\underset{NH_2}{CH}-COOH \;+\; CHO$$

Pyridoxamine phosphate   Pyridoxal phosphate

Both aspartate and alanine transaminases are present in the serum and are useful in diagnosis of disease. A marked rise in the concentration of aspartate transaminase in the serum is indicative of myocardial infarction, a heart condition involving the cardiac muscle, whereas an increased concentration of alanine transaminase is seen in viral hepatitis. Transamination reactions serve as important links joining carbohydrate, fat, and protein metabolism. A keto acid from any source can be used for the synthesis of an amino acid to be incorporated in tissue protein. For example, α-ketoglutaric acid is an intermediate in the Krebs cycle; pyruvic acid is an intermediate in the Embden-Meyerhof pathway. They both serve as links between protein metabolism and carbohydrate and lipid metabolism.

Formation of Urea

The ammonia, carbon dioxide, and water that result from the deamination and oxidation of the amino acids are combined to form urea. Urea formation takes place in the liver by a fairly complicated series of reactions, first described by Krebs and his coworkers. The ammonia and carbon dioxide combine with the amino acid ornithine to form another amino acid, citrulline. Another molecule of ammonia then combines with the citrulline to form the amino acid arginine, which is then hydrolyzed by means of the enzyme arginase, present in the liver, to form urea and ornithine. The ornithine may then enter the beginning of the cycle and combine with more ammonia and carbon dioxide from protein catabolism.

In recent years the detailed mechanism of the cycle has been worked out. Apparently ornithine does not react directly with CO_2 and NH_3 to form citrulline, but reacts with a compound called **carbamyl phosphate.** This compound is synthesized from ATP, CO_2, and NH_3 in the presence of the specific enzyme carbamyl phosphate synthetase and the cofactors N-acetylglutamate and Mg^{+2}.

$$CO_2 + NH_3 + 2ATP \xrightarrow[Mg^{+2},\ N\text{-acetylglutamate}]{\substack{\text{carbamyl}\\ \text{phosphate synthetase}}} NH_2-\underset{\underset{O^-}{|}}{\overset{\overset{O}{\|}}{C}}-O-\overset{\overset{O}{\|}}{P}-OH + 2ADP + P_i$$

Carbamyl phosphate

The use of two molecules of high-energy ATP in the formation of carbamyl phosphate apparently is required to make the reaction irreversible and to maintain the concentration of NH_3 in tissues and body fluids at a very low level. The formation of arginine is also not a simple reaction of citrulline and NH_3 but involves a combination with aspartic acid to form argininosuccinic acid, which then splits into arginine and fumaric acid. The currently accepted **urea cycle** can be represented as follows:

One of the nitrogen atoms and the carbon atom of urea come from the NH_3 and CO_2 (shaded). The other nitrogen atom comes from aspartic acid.

As urea is formed in the liver it is removed by the blood stream, carried to the kidneys, and excreted in the urine. Urea is the main end product of protein catabolism and accounts for 80 to 90 per cent of the nitrogen that is excreted in the urine.

NUCLEOPROTEIN METABOLISM

In Chapter 3 nucleoproteins were shown to be constituents of nuclear tissue composed of a protein conjugated with nucleic acids. The important nucleic acids DNA and RNA are essential constituents of the cell nucleus, the chromosomes, and viruses and are involved in the synthesis of protein. During the process of digestion the protein is split from the nucleic acids and is broken down to amino acids. The nucleic acids are first attacked by ribonuclease and deoxyribonuclease to form nucleotides that are further hydrolyzed by nucleotidases to form phosphates and nucleosides. The nucleosides are absorbed through the intestinal mucosa and split by nucleosidases of the tissues into

D-ribose, deoxyribose, purines, and pyrimidines. In metabolism the amino acids and sugar follow the ordinary process of protein and carbohydrate utilization. The phosphoric acid is used to form other phosphorus compounds in the body or may be excreted in the urine as phosphates.

Purine Metabolism

The synthesis of purines has been elaborated in recent years. This is a very complex process involving several steps, specific enzymes, and cofactors. Inosinic acid in the form of a mononucleotide is formed first and serves as an intermediate in the synthesis of adenylic acid and guanylic acid. With the help of tagged molecules, the precursors of the purine nucleus were established. These may be represented in the following diagram:

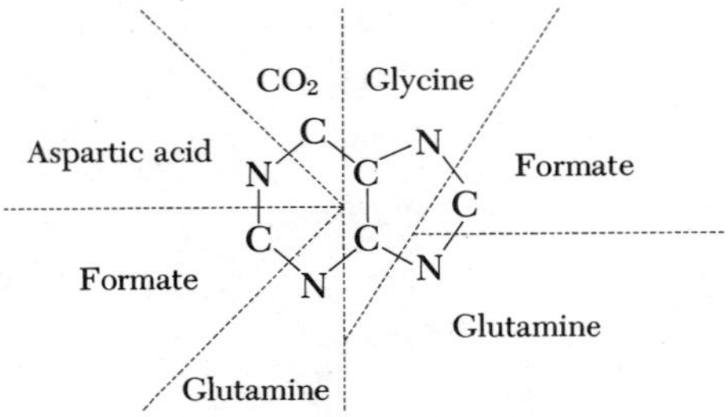

The purines that are formed by the hydrolysis of nucleosides in the tissues undergo catabolic changes, forming uric acid, which is excreted in the urine. The nucleosides **adenosine, inosine, guanosine,** and **xanthosine** are split into ribose plus adenine, hypoxanthine (6-oxypurine), guanine, and xanthine (2,6-dioxypurine), respectively. These purines are not completely broken down to NH_3, CO_2, H_2O, and energy, but are progressively oxidized with the assistance of specific enzymes.

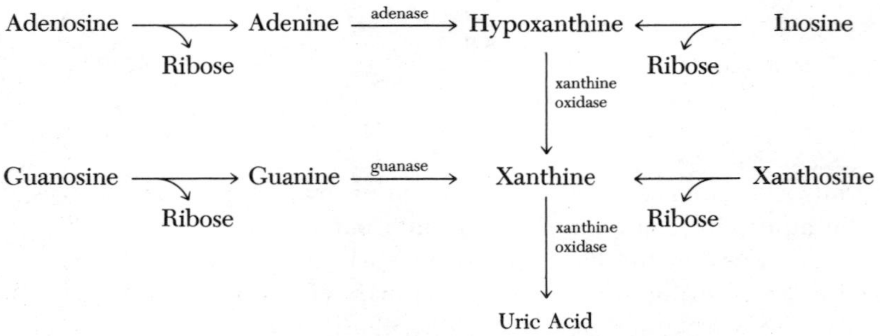

In most mammals other than man and apes the uric acid is converted into **allantoin,** a more soluble substance.

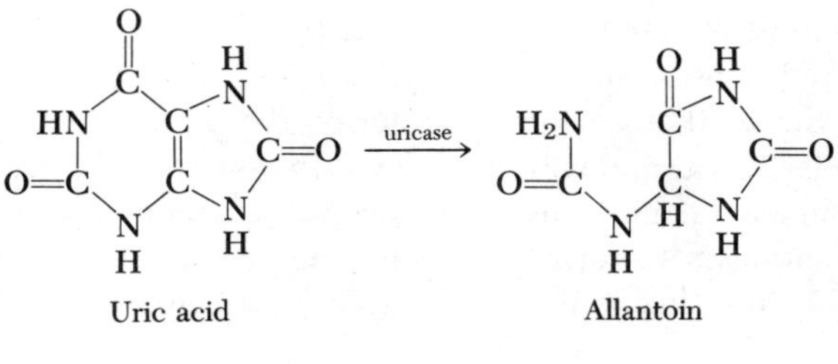

Uric acid   Allantoin

Pyrimidine Metabolism

Pyrimidine synthesis starts with ammonia and carbon dioxide reacting with ATP to form **carbamyl phosphate** (see p. 154). This compound combines with aspartic acid to form carbamyl aspartic acid, which is converted to dihydroorotic acid, which is reduced to orotic acid (6-carboxyl uracil). Orotic acid then reacts with 5-phosphoribosyl-1-pyrophosphate to form the nucleotide orotidine-5-phosphate. Decarboxylation of this nucleotide produces the primary pyrimidine, uridine-5-phosphate, or uridylic acid. The mononucleotide of uridylic acid apparently serves as the starting material for the synthesis of other pyrimidine nucleotides.

The pyrimidines that result from nucleoside hydrolysis in the tissues can be broken into small molecules in catabolism. Cytosine loses ammonia to form uracil, which is reduced to dihydrouracil utilizing the coenzyme NADPH. The ring is then opened to form a ureidopropionic acid, which is further split to β-alanine plus ammonia and carbon dioxide. These end products are eventually converted to urea for excretion.

CREATINE AND CREATININE

Creatine and creatinine are two nitrogen-containing compounds that are usually associated with protein metabolism in the body. **Creatine** is widely distributed in all tissues but is especially abundant in muscle tissue, where it is combined with phosphoric acid as **phosphocreatine**, or **creatine phosphate**. In the contraction of muscles, phosphocreatine apparently plays an important role as a reservoir of high-energy phosphate bonds readily convertible to ATP. The energy for the initial stages of muscular contraction probably

comes from the hydrolysis of this compound to form creatine and phosphoric acid. These two substances are later combined during the recovery period of the muscle (see p. 133). Creatine is synthesized from the amino acids glycine, arginine, and methionine.

Creatinine is also present in the tissues but is found in much larger amounts in the urine. It is formed from either creatine phosphate or creatine and is an end product of creatine metabolism in muscle tissue.

CORRELATION OF CARBOHYDRATE, LIPID, AND PROTEIN METABOLISM

The correlation between carbohydrate and lipid metabolism has already been discussed. Since the catabolism of amino acids results in keto acids such as pyruvic acid, it can readily be seen that these products could enter the metabolic scheme of the carbohydrates. Furthermore, the glucogenic amino acids and the ketogenic amino acids could enter the carbohydrate and lipid metabolism schemes. The over-all correlation of the three major types of metabolism is represented in the following diagram:

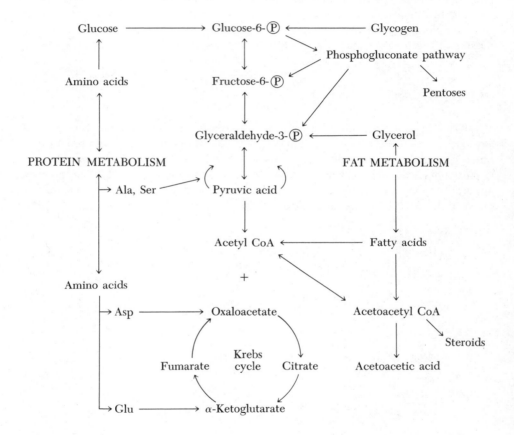

CARBOHYDRATE METABOLISM

IMPORTANT TERMS AND CONCEPTS

amino acid pool	m-RNA
anticodon	t-RNA
codon	transamination
creatinine	transcription
deamination	translation
essential amino acids	urea cycle
ribosomes	uric acid

QUESTIONS

1. Explain the concept of the amino acid pool.

2. What is meant by the dynamic state of tissue proteins in the body?

3. What is an essential amino acid? What would happen to a growing child that was deprived of adequate amounts of these amino acids? Why?

4. Describe the process of activation of amino acids and their combination with t-RNA prior to the synthesis of protein.

5. Discuss the properties and importance of t-RNA in protein synthesis.

6. Illustrate the process of translation in protein synthesis.

7. In protein synthesis, explain the relationship between codons and anticodons.

8. Briefly discuss the role of ribosomes in protein synthesis.

9. What are the processes of deamination and transamination? Illustrate one process with equations.

10. How does pyridoxal phosphate function as a coenzyme in the process of transamination?

11. Outline the essential reactions in the urea cycle.

12. What products would result from the complete hydrolysis of RNA? Briefly describe the metabolic fate of each of the products.

13. What compounds are involved in the synthesis of the purine nucleus?

14. Briefly outline the process for the synthesis of creatine.

15. Explain how the process of transamination can serve as a common link between carbohydrate, fat, and protein metabolism?

Chapter 13
THE BIOCHEMISTRY OF GENETICS

The *objectives* of this chapter are to enable the student to:

1. Explain why single cells such as bacteria are used in the study of genetics.
2. Explain the relation between DNA of the cell and genetics.
3. Describe the function of genes and their relation to the chromosomes of the cell.
4. Illustrate the process of replication of DNA.
5. Describe and illustrate the regulation of enzyme synthesis in the cell.
6. Recognize that inborn errors of metabolism are genetic diseases.
7. Outline the inborn errors of metabolism related to phenylalanine metabolism.
8. Discuss the possibility of genetic engineering for the reversal of cancer and repair of genetic disease.

In previous chapters the structure of the living cell, its subcellular particles, its chemical composition, and many of its chemical reactions have been considered. An essential role of the cell that has not been discussed is its role in the transmission of genetic information. The composition and structure of the RNA and DNA molecules of the cell were covered in Chapter 3 and the relation between DNA, chromosomes, and genes was suggested. In this chapter we will explore briefly the biochemistry of heredity and the genetic information transmitted by the DNA and RNA molecules.

CELLS AND HEREDITY

Most of the biochemical studies in genetics have been carried out with single cells such as bacteria, which are capable of producing many generations in a short period of time. Bacteria are simple systems that contain relatively little genetic information compared to the cells of man and animals. The number of genes in a bacterial cell would be very low, and there should be a small amount of DNA as compared to that in cells of higher organisms. These facts make bacteria ideal systems for research in genetics. Extensive biochemical research investigations have established the genetic makeup of these cells, and proposals for the regulation of the transcription of genetic information by the cells have been advanced. This type of research has developed into a new area of science called **molecular biology,** which is concerned with explaining genetics on a molecular level. It should be realized that knowledge about DNA and about RNA and

160

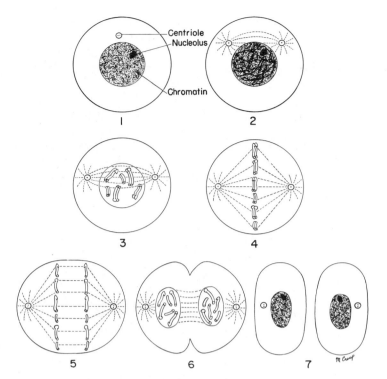

FIGURE 13-1   Mitosis in a cell of a hypothetical animal with a diploid number of six (haploid number = 3); one pair of chromosomes short, one pair long and hooked, and one pair long and knobbed. 1, Resting stage. 2, Early prophase: centriole divided and chromosomes appearing. 3, Later prophase: centrioles at poles, chromosomes shortened and visibly doubled. 4, Metaphase: chromosomes arranged on equator of spindle. 5, Anaphase: chromosomes migrating toward poles. 6, Telophase: nuclear membranes formed; chromosomes elongating; cytoplasmic divisions beginning. 7, Daughter cells: resting phase. (From Mazur and Harrow: Textbook of Biochemistry, 10th edition, Philadelphia, W. B. Saunders Company, 1971, p. 477.)

DNA replication occurring in single cells may not be directly applied to human genetics, but it greatly increases our understanding of the genetic processes in the body.

The nucleus of the cell, which contains most of the DNA (Chapter 1), is of prime importance in genetics. The **chromosomes,** which contain the genes, are thought to be large DNA molecules. A developing embryo contains **germ cells,** which are **haploid** and possess only one set of chromosomes, and **somatic cells,** which are **diploid** and possess two sets of chromosomes. The somatic cells give rise to the tissues and organs characteristic of the species, while the germ cells, at sexual maturity, are responsible for the development of eggs or sperm and thus for the passing of the genetic characteristics of the species. As the cells divide during the process of reproductive replication, the chromosomes and genes divide into exact duplicates of themselves which are finally deposited in the nucleus of each daughter cell, as shown in Figure 13–1. The chromosomes and genes of the germ cells and somatic cells, by this process of replication, transmit the genetic information from the parents to the offspring.

DNA AND GENETICS

The relation of DNA to genetics was first recognized in the 1940s. In 1943 Avery observed that the DNA from a virulent strain of bacteria transformed a nonvirulent strain of the same microorganism into a virulent strain. It is significant that the haploid cells contain only half the amount of DNA found in diploid somatic cells, and that in a given species the amount of DNA per diploid cell is fairly constant from one type of cell to another. Also, it has been shown that the purine and pyrimidine base composition of DNA varies from one species to another. The DNA in the cells or tissues of a single species has the same base composition and does not change with nutritional status, age, or a change in environment. As discussed in Chapter 3, the number of guanine bases is equal to the number of cytosine residues, and the number of adenine residues is equal to the number of thymine residues in all DNA's. In addition, the sum of the purine bases equals the sum of the pyrimidine bases, and the G-C and A-T pairs with their accompanying hydrogen bonds (Fig. 3–4) not only fit best in the double helix structure (Fig. 3–3), but also exhibit the strongest hydrogen bonding and stability.

CHROMOSOMES AND GENES

In experiments with bacteria and bacterial viruses, it has been demonstrated that single chromosomes of DNA-containing bacteria and viruses consist of single very large double helical DNA molecules. For example, the chromosome of a commonly studied bacteria, *E. coli*, consists of a single DNA molecule with a molecular weight of about 2.8 billion, about 4.2 million base pairs, and a contour length of over $1200\,\mu$. Cells of higher organisms contain several chromosomes, each of which consists of one or more large DNA molecules. The DNA in a single human cell has been calculated to contain 5.5 billion base pairs and to have a total length of about 2 meters.

Genes are segments of a chromosome that code for a single polypeptide chain of a protein or enzyme. Since the sequence of three nucleotides in DNA—a codon (p. 151)—is required to code for a single amino acid residue, the size of the gene which determines a specific protein may be calculated by multiplying the number of amino acid residues in the protein by 3. In proteins or enzymes that have two or more different polypeptide chains, each coded by a different gene, more than one gene may be required to code for these proteins. The exact location of a gene in the chromosome of a bacterial cell that codes for a specific protein can be established by complex genetic mapping methods that are beyond the scope of this book. Fortunately, the genes that code for the individual enzymes of a multienzyme system are often located adjacent to each other in the chromosomes and are transcribed and translated as a group. The single chromosome of bacteria or bacterial viruses may contain from 5 to as high as 4000 different genes.

REPLICATION OF DNA

The mechanism by which genetic information can be accurately replicated was proposed by Watson and Crick, based on their double helical structure of DNA. Since the two strands of DNA are structurally complementary to each other and therefore contain complementary information in their base sequences, the **replication** of DNA during cell division was suggested to occur by separation of the two strands, each becoming a template to specify the base sequence of a new complementary strand. The enzyme **DNA polymerase**, first extracted from bacteria (*E. coli*), catalyzes the replication. An illustration of this process with the formation of two daughter double helical DNA's, each containing one strand from the parent DNA, is shown in Figure 13–2.

In reduplication of cell nuclei, which is necessary in cell division, the double helix may unravel, and each of the original chains may serve as a template for the synthesis of another chain. It has been shown experimentally that by adding labeled purine and

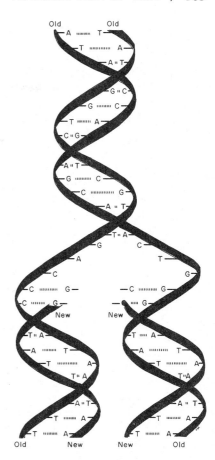

FIGURE 13-2   Replication of DNA as suggested by Watson and Crick. The complementary strands are separated and each forms the template for synthesis of a complementary daughter strand. (From Villee: Biology. 6th ed. Philadelphia: W. B. Saunders Company, 1977.)

pyrimidine intermediates to a synthesizing system one chain of newly synthesized DNA contains labeled intermediates from the system, whereas the original chain is unlabeled. Since the base-pairing pattern of DNA is followed, the newly synthesized chain will possess the exact nucleotide sequence of the original parent chain. The result is the synthesis of two pairs of DNA chains (Fig. 13–2) in which each pair is identical in nucleotide sequence and genetic coding information to the original pair of parent chains. In the laboratory it has been demonstrated that pure DNA preparations from a particular species of bacteria or bacteriophage, when added to another species of bacteria, will serve as a template to direct the recipient cells to develop the characteristics of the donor bacteria. It is quite possible that the multiplication of viruses within cells may occur by the same process. For example, type 1 poliomyelitis virus, which has been crystallized, contains an RNA that serves as a template to infect cells with this type of virus. This knowledge suggests that RNA functions mainly in the cytoplasm of a cell as a template for the synthesis of specific cellular proteins. A close relation exists between DNA of the nucleus and RNA of the cytoplasm, since one chain of DNA and one chain of RNA could twist around each other to form a double helix and thus influence the RNA template.

TRANSCRIPTION OF DNA

The replication of DNA in the nucleus of cells to preserve and pass on its genetic information to the DNA of daughter cells is obviously of prime importance in genetics.

The next step, that of directing protein-synthesizing machinery to produce specific molecules such as enzymes and hormones required in metabolic reactions in the body, is called **transcription** (p. 150). In earlier studies of protein synthesis in intact cells, Jacob and Monod observed an increased rate of synthesis of cytoplasmic RNA. They proposed that this species of RNA served as a messenger carrying genetic information from the DNA of the chromosomes to the surface of the ribosomes. In addition, they suggested that this **messenger RNA** is formed enzymatically so that it has a base sequence complementary to that of one strand of RNA. The m-RNA was believed to contain the complete message necessary for specifying one or more polypeptide chains and for binding the ribosomes to serve as a functioning template for protein synthesis (p. 150).

The discovery of DNA polymerase that was required for DNA replication stimulated a search for **RNA polymerase,** the catalyst for DNA transcription. This enzyme was also obtained from extracts of *E. coli*. Messenger RNA synthesis was found to require RNA polymerase, ribonucleoside triphosphates and double-stranded DNA molecules. Only one strand of the DNA is transcribed as it imparts its genetic message to m-RNA (Fig. 12–2).

REGULATION OF PROTEIN SYNTHESIS

After the DNA of the chromosomes imparts its encoding message to the m-RNA and this molecule directs the synthesis of specific protein molecules on the ribosomes, what process directs the amount and type of proteins produced? It seems obvious that the living cell must possess a mechanism for regulating the relative amounts of the different protein molecules that are synthesized. Enzymes, for example, that are required in the main metabolic cycles in the cells are necessarily synthesized in greater quantities than those needed for the synthesis of coenzymes. There must also be a mechanism for turning on or off the synthesis of proteins as they are needed for cellular functions.

Most of our information about regulation comes from studies on the control of synthesis of enzyme proteins. Jacob and Monod postulated that the chromosomes carry three types of genes, **structural genes, operator genes,** and **regulator genes.** The structural gene DNA directs the synthesis of protein molecules as described previously. The operator gene controls the action of adjacent structural genes in the synthesis of specific m-RNA molecules. The structural genes and their operator gene are designated an **operon.** The

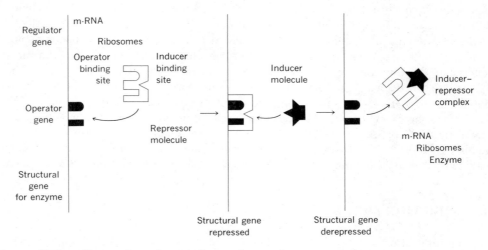

FIGURE 13–3   Proposed mechanism for regulation of enzyme synthesis.

operator gene itself is controlled by a regulator gene which directs the synthesis of protein molecules called **repressors.** When the repressor combines with its operator gene, the structural genes cannot function in protein synthesis, and the operon is said to be repressed. If a situation in the cell inactivates the repressor and permits the operon to function, the operon is said to be derepressed (Fig. 13–3). As suggested in Chapter 4, the end products of enzyme reactions, which are often metabolites, can affect the activity of enzyme systems. They may carry out this function by control of the repressor and the enzyme synthesis mechanism described previously.

THE CENTRAL DOGMA OF GENETICS

Since about 1955, a major objective in biochemistry has been to document and clarify the details of what is called the **central dogma of genetics:** that DNA is the hereditary material; that its information is encoded in the sequences of its subunits, the genes; and that this information is transcribed onto RNA and then translated into protein. Recently, in studies of the bacterial genes of E. coli, electron micrographs were obtained that show both an inactive chromosome segment and an active chromosome segment. In the active segment the DNA is seen being transcribed onto messenger RNA and the RNA being translated into protein. The electron micrographs obtained in these studies bear a striking resemblance to diagrams of transcription and translation that have been proposed by recognized research workers in this area.

In another recent study of how cells proceed through division in the cell cycle, an electron micrograph clearly depicting chromosome replication was obtained. The DNA double helix does not unwind as neatly as shown in Figure 13–2 but appears as tangled strands with segments of replicating DNA molecules joining the original DNA strand at several junctures in the micrograph.

Multiplication of viruses within cells can occur by a similar process in which the viral genes consist of DNA that transmits information to the RNA of the cell and then into the cell proteins. Several common viruses which cause poliomyelitis, the common cold, and influenza are called **RNA viruses,** since the RNA replicates directly into new copies of RNA and translates information directly to proteins of the cell without DNA involvement in their replication. Recently evidence for a reverse flow of genetic information from RNA to DNA has been obtained in experiments using the Rous sarcoma virus. It has been suggested that in normal cells there are regions of DNA that serve as templates for the synthesis of RNA, and that this RNA serves in turn as a template for the synthesis of DNA which then becomes integrated with the cellular DNA. These experiments may be valuable in an explanation of the origin of cancer in humans. Since it was shown that cancer-causing RNA viruses can produce a DNA transcript of the viral RNA, it is possible that the viral RNA may transmit genetic information to the genes that will eventually surface as spontaneous cancer.

TOPIC OF CURRENT INTEREST

GENETIC ENGINEERING

Investigators in many scientific disciplines have combined their efforts in an attempt to understand the workings of the genetic machinery of the living cell. In the past two decades a marked increase in this research has been carried out by specialists called molecular geneticists. They have provided information about the structure of chromosomes, genes, plasmids, DNA, and smaller molecules involved in the transfer of genetic information. In addition, they have isolated and purified enzymes such as DNA polymerase, DNA ligases, terminal transferase, and restriction endonucleases, which serve as tools in the alteration of the genetic machinery of the cell, and studied their properties.

The series of investigations, which led to successful manipulation of genes, began in 1967 with reports from several laboratories of the discovery of **DNA ligases.** These are enzymes that can join together the loose ends of DNA strands that have complementary blocks of nucleotides on the 3′ and 5′ ends of the strand. This discovery was followed by the development of enzymes that recognized specific sites on DNA molecules and cleaved them at those sites. In the early 1970s, the nucleotide sequences at the cleavage sites recognized by several of the enzymes, called **restriction endonucleases,** were identified. One of these enzymes, named Eco RI enzyme, was isolated in Boyer's laboratory at the University of California at San Francisco and was found to produce breaks in two DNA strands that were separated by several nucleotides. Because of the symmetrical arrangement of nucleotides in the region of cleavage, the two split strands yielded DNA termini with complementary nucleotide sequences that acted as "sticky" mortise and tenon ends. In 1972, Mertz and Davis of Stanford University used Eco RI enzyme to cleave a closed-loop bacterial virus DNA and found that the virus DNA molecules would re-form themselves into circular molecules by hydrogen bonding and could be sealed with the enzyme DNA ligase. These investigators were also able to demonstrate that the reconstituted virus DNA molecules were infectious in animal cells growing in tissue culture. This was a major step in devising a system to alter genes. However, most segments of DNA are not self-replicating and need to be integrated into DNA molecules which can replicate in a particular system. An excellent candidate for this phase of genetic manipulation is a **plasmid,** a small DNA molecule that exists apart from the chromosomes in bacteria and often carries genes for some supplementary activity, such as resistance to antibiotics. Cohen and his group at Stanford separated from *E. coli* chromosomes a small plasmid, which had genes for multiple antibiotic resistance. After treating the plasmid DNA with Eco RI enzyme, they found by electrophoretic analysis that the plasmid molecule was cut in only one place, producing a single linear fragment. They joined the ends of the fragment by hydrogen bonding and sealed them with DNA ligase. When they introduced the reconstituted circular molecules into *E. coli,* they were biologically functional plasmids that replicated and still conferred tetracycline resistance. These experiments were followed by a successful attempt to insert at the cleavage site in the tetracycline-resistant plasmid a portion of an *E. coli* plasmid that carried resistance to the antibiotic kanamycin. The plasmids isolated from these studies were found to contain the entire original plasmid plus a second DNA fragment that carried the gene for kanamycin resistance.

A very important next step was to determine if genes from other species could be introduced into *E. coli* plasmids. Cohen and Chang, utilizing the techniques described above, produced transformed bacteria from a mixture of *E. coli* and *S. aureus* that contained a new DNA species that had both the characteristics of the staphylococcal and coliform plasmid DNA. Several investigators, including Cohen, Boyer, Chang, Morrow, and Goodman, then combined their efforts to determine whether animal cell genes could be introduced into bacteria, and whether they would replicate in the transformed bacteria. This research resulted in a successful transplant of highly purified ribosomal genes of a toad, *Xenopus laevis,* into the plasmid DNA of the bacterium *E. coli.* The animal cell genes were reproduced in several generations of the bacteria, and, in addition, the nucleotide sequences of the toad DNA were transcribed into an RNA product in the bacterial cells. Since 1973, when these experiments were reported, similar experiments in the replicating or **cloning** of DNA have been carried out in several laboratories. Some workers have isolated from complex chromosomes certain regions that are implicated in particular functions such as replication. It has also been possible to isolate groups of genes that are active at a specific stage in an animal's development.

The manipulation of genes offers many exciting possibilities. Bacterial cells might be constructed that could be grown easily and inexpensively, and which would synthesize a variety of biologically-produced substances such as antibiotics, vitamins, or hormones in quantities that will compare acceptably with those produced by methods currently used by the pharmaceutical industry. It is conceivable that by introducing genes which increase the efficiency of photosynthesis, crop production could be improved and the nutritive value of plant products increased. In addition, the world's food supply might be increased by introducing the nitrogen fixation system of bacteria into plants or a symbiotic organism. In medical research, gene manipulation could be used as therapy in the treatment of diseases with a genetic basis.

The major question at present is how to control genetic engineering experiments in the future, since uncontrolled research might result in the propagation of infectious viruses, unknown diseases, and unwanted human mutations. In the spring of 1974, the National Institutes of Health and the National Science Foundation organized a meeting of investigators from the United States and 16 other countries to consider aspects of the control of these experiments. They discussed the hazards, safety factors, and proper scientific protocol for future experiments. They agreed that the plasmid or viral DNA into which foreign DNA has been inserted should be able to propagate only in specified bacterial hosts and only under defined conditions, and that the bacterial hosts should be unable to survive outside the laboratory environments. If proper control can be exercised in gene manipulations, then genetic engineering offers many exciting possibilities for improving life in the world; if not, it could result in many unknown and unwanted problems.

INBORN ERRORS OF METABOLISM

As early as 1906, Garrod, an English physician, described several abnormal patterns of metabolism. He called these abnormalities **inborn errors of metabolism,** since he recognized that the conditions were inherited. To the present time, approximately one hundred of these genetic diseases have been reported. Many of them are extremely rare. The diseases due to a metabolic block in amino acid catabolism are of special interest since they occur somewhat more frequently. A common clinical manifestation of such conditions is mental deficiency. Three of these inborn errors, **phenylketonuria, alkaptonuria,** and **tyrosinemia,** will be considered in more detail as they are concerned with the metabolism of phenylalanine, an essential amino acid. Each involves a deficiency of a different single, specific enzyme whose synthesis is genetically controlled.

Phenylketonuria

Phenylketonuria, or PKU, was first recognized by Fölling in 1934 when he detected large amounts of **phenylpyruvic acid** in the urine of several mentally retarded patients. All individuals who have so far been found to excrete phenylpyruvic acid in their urine daily have shown some degree of mental deficiency; about 1 per cent of the patients in institutions for the mentally retarded excrete phenylpyruvic acid in their urine. Since the condition was recognized as a genetic disease, attention has been focused on parents as carriers of the defect and on detection and treatment of newborn infants with the disease.

The normal metabolism of phenylalanine involves its transformation into tyrosine with the aid of **phenylalanine hydroxylase.** If sufficient quantities of this enzyme are not synthesized, as in phenylketonuria, the concentration of phenylalanine in the blood, spinal fluid, and urine increases. In the tissues, the phenylalanine is converted into phenylpyruvic, phenyllactic, and phenylacetic acids as shown in the following scheme. These metabolites of phenylalanine are excreted in the urine in large amounts in phenylketonuria.

As the disease itself is rather well understood today, the practical problem consists of identifying infants with the disease and initiating adequate treatment. Simple tests for the detection of phenylpyruvic acid in the urine or, preferably, tests for the increased concentrations of phenylalanine in the blood are available and are required in several

Phenylalanine → Tyrosine (phenylalanine hydroxylase); Phenylpyruvic acid, Phenyllactic acid, Phenylacetic acid

states. The most successful treatment consists of restricting the amount of phenylalanine in the diet of the PKU infant, providing only the minimum quantity essential for normal growth and development. Fortunately, this regimen allows the child to develop normally; the diet may apparently be relaxed somewhat after the age of six without serious effects on the child.

Alkaptonuria

Alkaptonuria is a rare genetic abnormality of tyrosine metabolism that is readily recognized by the characteristic changes in color which occur in the urine. When freshly passed the color is normal, but on standing it begins to darken. Alkalinity speeds up the change, and the urine passes through shades of brown to a final black color. Diapers wet with urine become darkly stained and as a result the condition is frequently recognized in early infancy. In adults, alkaptonuria may first be detected in a life insurance examination

Tyrosine → p-Hydroxyphenylpyruvic acid → Homogentisic acid (p-hydroxyphenylpyruvate oxidase) → Fumaric acid + Acetoacetic acid (homogentisic acid oxidase)

or routine medical check. The urine has strong reducing properties, which gives rise to a positive Benedict's test. These characteristics persist throughout life. As patients grow older their ligaments and cartilages tend to become dark blue in color, and the patients are prone to develop osteoarthritis.

The normal metabolism of tyrosine, which may be formed from phenylalanine as previously discussed, includes the formation of **homogentisic acid,** which is oxidized to fumaric acid and finally to acetoacetic acid. This oxidation is catalyzed by the enzyme **homogentisic acid oxidase.** The metabolic error is a block in the breakdown of the homogentisic acid, caused by a decrease in the concentration of the oxidase in the tissues; normal individuals can readily oxidize increased quantities of this acid without producing pigmented urine specimens. The formation of homogentisic acid and its metabolites is shown in the scheme on the opposite page. At present there is no specific treatment for this rare genetic disease.

Tyrosinemia

Tyrosinemia is a rare inborn error of tyrosine metabolism seen primarily in premature infants. In the scheme on the opposite page, the metabolic block associated with this defect occurs between p-hydroxyphenylpyruvic acid and homogentisic acid. The enzyme required for this step, **p-hydroxyphenylpyruvate oxidase,** has been found to be in low concentration in the liver of premature infants. In infantile tyrosinemia, kidney defects and a nodular cirrhosis of the liver occur, causing rickets, thrombocytopenia, darkening of the skin, and slight mental retardation. The urine contains large amounts of tyrosine and metabolites, including p-hydroxyphenylpyruvic, p-hydroxyphenylacetic, and p-hydroxy-phenyllactic acids. The disease may be acute or chronic; the acute cases are characterized by diarrhea, failure to thrive, and death from liver failure in the first seven months of life.

Ascorbic acid is the coenzyme of p-hydroxyphenylpyruvate oxidase, and the disease may be alleviated in premature infants by feeding excess ascorbic acid. Longer term treatment for tyrosinemia consists of a diet low in tyrosine and phenylalanine. If the dietary treatment is started early in life, both liver and kidney damage may be prevented.

IMPORTANT TERMS AND CONCEPTS

alkaptonuria	inborn errors of metabolism
chromosomes	operon
DNA polymerase	phenylketonuria
genes	replication
genetic engineering	repressor
germ cells	somatic cells

QUESTIONS

1. What features of single cells such as bacteria are useful in the study of genetics?

2. Describe the characteristics of germ cells and somatic cells.

3. What evidence suggested that DNA is involved in genetic processes?

4. Discuss the close relationship between genes and chromosomes of the cell.

5. Explain with an illustration the process of replication of DNA.

6. Describe the function of DNA polymerase and RNA polymerase in replication and transcription of DNA.

7. Outline a mechanism for the regulation of enzyme synthesis in the cell.

8. What is the central dogma of genetics? Explain.

9. Why is the detection of phenylketonuria so important in newborn infants?

10. Briefly discuss the condition of alkaptonuria.

11. What is genetic engineering and how may it be of value in the future?

Chapter 14

BODY FLUIDS

The *objectives* of this chapter are to enable the student to:

1. Describe the process of normal distribution of fluids between the circulatory system and the interstitial spaces.
2. Recognize the factors involved in normal water balance.
3. Describe the essential components of plasma or serum and the separation of proteins by electrophoresis.
4. Illustrate the normal composition of plasma electrolytes.
5. Describe the factors involved in the acid-base balance of the blood.
6. Describe the chemistry of hemoglobin and the changes that occur in the molecule in genetic disorders.
7. Recognize the essential functions of the kidney and the process of urine formation.
8. Describe the regulation of water, electrolyte, and acid-base balance by the kidney.
9. Describe the processes of H^+ formation and ammonia synthesis and their importance in kidney function.

    The composition of the fluid inside a living cell is to a considerable extent dependent on the fluid surrounding the cell. The extracellular fluid must carry nutrient material to the cell and waste products of metabolism away from the cell. The constant interchange of substances between the intracellular and extracellular fluid depends on the process of diffusion and on transport mechanisms that are often driven by the energy of ATP. The fluid balance between tissues and organs requires a circulatory system of greater magnitude than that of the cell. The blood, for example, transports end products of digestion from the intestine to the liver, where they are processed and then carried to other tissues and organs and finally, as waste products of metabolism and excess mineral salts, are transported to the kidneys for excretion. The important gases, oxygen and carbon dioxide, as well as enzymes, hormones, and other regulatory substances, are also carried in the blood.

    The **blood** is the most active transport system and consists of cellular elements suspended in plasma, the liquid medium. The **tissue fluid,** which surrounds the tissues, and the **lymph,** a slow-moving fluid that is similar to plasma and is carried in a system of vessels called the **lymphatics,** also assist the transportation system of the body. Lymph and tissue fluid are known together as **interstitial fluid** and make up about 15 per cent of the body's weight. Important interchange of material occurs from the blood to the tissue through the medium of the interstitial fluid. Plasma and interstitial fluid are collectively considered as **extracellular fluids.** A 160-pound man would have about 6.0

liters of blood, 3.5 liters of plasma, 10.5 liters of interstitial fluid, and 35 liters of intra-cellular fluid.

The intracellular fluids are separated from the extracellular fluids by the cellular membranes. Water, electrolytes, nutrient material, and waste products must pass through the membrane to maintain cellular function. The cell membrane is freely permeable to water, O_2, CO_2, urea, and glucose, but exhibits a selective permeability toward electrolytes, such as Na^+, K^+, Cl^-, Ca^{+2}, Mg^{+2}, and HCO_3^-. In experiments during which heavy water, D_2O, was injected into the extracellular fluid, it was found that the body required about 120 minutes to establish equilibrium between the intra- and extracellular fluids.

Water is rapidly transported throughout the body by the circulating plasma of the blood. The movement of water between the plasma and the interstitial spaces is dependent on the proteins in the plasma, which constitute a colloidal-osmotic machine that can draw water into the blood capillaries. This osmotic pressure, called the **oncotic pressure,** is equivalent to 23 torr and is opposed by the hydrostatic pressure transmitted to the capillaries from the heart. This pressure averages 32 torr at the arterial end and 12 torr at the venous end of the capillary. At the arterial end of the capillary system the hydrostatic pressure forces fluid into the interstitial space, whereas at the venous end the colloid osmotic pressure pulls it back into the capillaries. This mechanism is responsible for the normal distribution of fluids between the blood stream and the interstitial space.

A decrease in plasma protein, which may occur in serious malnutrition or in kidney disease, would decrease the oncotic pressure and result in an increased flow of fluid to the interstitial space. This condition is known as **edema,** and it can also occur in any heart disease that results in an increase in blood pressure.

If *water balance* is to be maintained in the body, it is obvious that fluid intake and fluid excretion must be equal. The intake and output of a normal adult of average size can be represented as in the following tabulation:

Intake	ml/Day	Output	ml/Day
		Insensible perspiration	
		Lungs	700
		Skin	300
Water in beverages	1200	Sweat	300
Water in food	1500	Feces	200
Water of oxidation	300	Urine	1500
	3000		3000

Some concept of the magnitude of the daily fluid turnover by the body can be gained from the following data:

Daily Volume of Digestive Fluids

	ml/Day
Saliva	1,500
Gastric secretion	2,500
Intestinal secretion	3,000
Pancreatic secretion	700
Bile	500
	8,200
Fluids lost from the body	3,000
Total fluid turnover	11,200
Compared to:	
Plasma volume	3,500
Total extracellular fluid	14,000

In general, if the fluid intake exceeds the output for any length of time, the tissue fluid will increase in volume and **edema** will result. The excessive loss of body fluids that may occur from vomiting, diarrhea, or copious sweating causes **dehydration** of the body.

BLOOD

Some idea of the importance of blood to the body can be gained from a consideration of its major functions:

1. The blood transports nutrient material to the tissues and waste products of metabolism to the organs of excretion.
2. It functions in respiration by carrying oxygen to the tissues and carbon dioxide back to the lungs.
3. It distributes regulatory substances, such as hormones, vitamins, and certain enzymes, to the tissues in which they exert their action.
4. The blood contains white corpuscles, antitoxins, precipitins, and so on, which serve to protect the body against microorganisms.
5. It plays an important role in the maintenance of a fairly constant body temperature.
6. It aids in the maintenance of acid-base balance and water balance.
7. It contains a clotting mechanism that protects against hemorrhage.

Blood comprises approximately 8 per cent of the body weight, which means that an average person has 4000 to 6000 ml of blood. Loss of blood by bleeding or by donating for transfusion has no serious effect on the body, since the blood volume is rapidly regenerated.

BLOOD CELLS

The two major portions of blood are the blood cells and the plasma. When separated by centrifugation, the blood cells occupy from 40 to 45 per cent by volume of the blood. This fraction contains the red blood cells, white blood cells, and thrombocytes (platelets).

The red blood cells, or **erythrocytes,** contain the respiratory pigment hemoglobin, and have several important functions. The number of red blood cells in men is approximately 5,000,000 per cubic millimeter, and in women approximately 4,500,000. The determination of the number of red cells in the blood is often carried out in the laboratory, since the value changes markedly in diseases such as anemia.

White blood cells, or **leukocytes,** are larger than red cells and have nuclei, which red cells do not. Normally, there are from 5000 to 10,000 white cells per cubic millimeter of blood. There are several types of white blood cells, and they all function to combat infectious bacteria. White cell counts are routinely determined in the laboratory since they increase in acute infections such as acute appendicitis.

Thrombocytes (platelets) are even smaller than red blood cells and do not contain a nucleus. There are from 250,000 to 400,000 thrombocytes per cubic millimeter in the blood of a normal person. Their major function is in the process of blood clotting, since they contain cephalin, a phospholipid involved in the early stages of clotting.

SERUM AND PLASMA

When freshly drawn blood is allowed to stand, it clots, and a pale yellow fluid soon separates from the clotted material. This fluid is called **serum** and is blood minus the

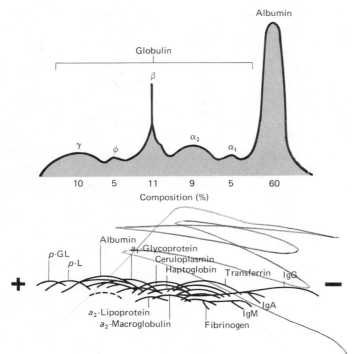

Figure 14-1 *Upper:* An electrophoretic pattern of normal plasma. *Lower:* Composite diagram of immunoelectrophoretic analysis of human plasma proteins. The anode is to the left. "Prealbumins" are labeled *p*-GL (for glycoprotein) and *p*-L (for lipoprotein). The pattern as drawn suggests the approximate relative positions of the more frequently observed precipitation arcs.

formed elements and fibrinogen, which is used in the clotting process. If, on the other hand, blood collected in the presence of an anticoagulant is centrifuged, the fluid portion that separates from the cells is called **plasma.**

Plasma Proteins

The proteins of the plasma are present in a concentration of about 7 per cent. The most important of these are **albumin,** the **globulins,** and **fibrinogen.** The globulins have been separated into several fractions of different molecular size and properties by the technique of **electrophoresis** (Fig. 14–1). More recent techniques, including column, starch gel and cyanogum electrophoresis, and immunoelectrophoresis, are capable of separating plasma into more than 20 different proteins.

The plasma proteins have several functions in the body. One of the most essential is the maintenance of the effective osmotic pressure of the blood, which controls the water balance of the body. The globulins, IgG, IgA, and IgM, contain immunologically active antibodies against such diseases as diphtheria, influenza, mumps, and measles. The lipoproteins are combinations of lipids and α or β globulins, and function in the transport of lipids. Transferrin binds iron and ceruloplasmin binds copper, and these proteins function in the transport of these metals. Fibrinogen is an essential component in the blood clotting process.

SERUM ENZYMES

There are several enzymes in the plasma or serum that are released into the blood by the breakdown of body tissues. Since marked changes from the normal concentration often occur in disease, assays for specific enzymes are currently a powerful diagnostic tool in the laboratory. Clinically significant enzymes are amylase, lipase, acid and alkaline phosphatase, lactic dehydrogenase (LDH), creatine phosphokinase (CPK), aldolase, aspartate transaminase (AST), and alanine transaminase (ALT). Amylase, for example, increases in acute pancreatitis, alkaline phosphatase in obstructive jaundice of the liver, and AST in myocardial infarction.

Plasma Electrolytes

The electrolytes that are present in the body fluids consist of positively charged ions, or **cations,** and negatively charged ions, or **anions.** They are mainly responsible for the osmotic pressure of the fluids and are involved in maintenance of the acid-base and water balance of the body. The major cations in the body fluids are Na^+, K^+, Ca^{+2}, and Mg^{+2}, whereas the major anions are HCO_3^-, Cl^-, HPO_4^{-2}, SO_4^{-2}, organic acids, and protein. In a consideration of electrolyte balance, the concentration of these ions is expressed in milliequivalents per liter of body fluid. Milliequivalents per liter (mEq/l) equals the atomic weight expressed in milligrams per liter divided by the valence. The concentration of plasma electrolytes as determined in the laboratory is sometimes expressed in mg/100 ml plasma. Calcium is often reported as 10.0 mg/100 ml for a normal individual, and the normal sodium concentration is about 327 mg/100 ml. To express these concentrations in mEq/l, we would proceed as follows:

$$\frac{mg/100 \text{ ml} \times 10}{\dfrac{\text{atomic wt.}}{\text{valence}}} = \frac{mg/l}{\text{equiv. wt.}} = mEq/l$$

$$\frac{10 \text{ mg}/100 \text{ ml} \times 10}{\dfrac{40}{2}} = \frac{100}{20} = 5 \text{ mEq/l of Ca}$$

$$\frac{327 \text{ mg}/100 \text{ ml} \times 10}{\dfrac{23}{1}} = \frac{3270}{23} = 142 \text{ mEq/l Na}$$

The electrolyte balance of the ions in the plasma is shown in the following tabulation:

Cations	mEq/l	Anions	mEq/l
Na^+	142	HCO_3^-	27
K^+	5	Cl^-	103
Ca^{+2}	5	HPO_4^{-2}	2
Mg^{+2}	3	SO_4^{-2}	1
		Organic acids	6
		Protein	16
Total	155	Total	155

Comparisons of the electrolyte composition of body fluids and changes in disease are often illustrated in vertical bar graphs devised by Gamble. Variations in the electrolyte content of normal plasma, interstitial fluid, and intracellular fluid are shown in Figure 14–2.

In the discussion of water balance we listed the volumes of the daily secretions of digestive fluids. Their electrolyte composition must also be taken into account, especially when they are lost from the body by vomiting or diarrhea. In comparing the electrolytes of these fluids with those of plasma, only the major components that are involved in acid-base or electrolyte balance are shown in the bar graphs in Figure 14–3. It is readily apparent that the digestive secretions vary in their composition of electrolytes and in their contribution to the acid-base balance of the body.

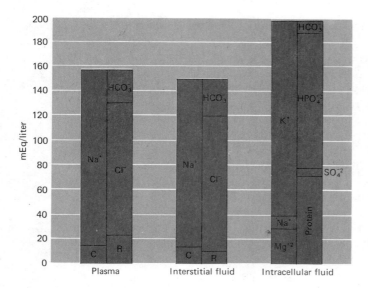

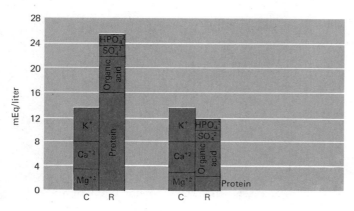

FIGURE 14-2   Electrolyte composition of normal body fluids. (The expanded scales below show the individual electrolytes in the C and R spaces.)

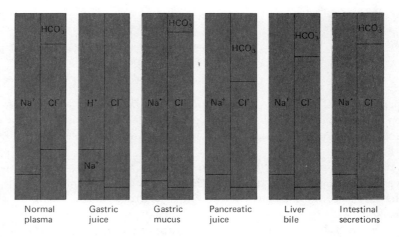

FIGURE 14-3   A comparison of the major electrolyte components in normal digestive secretions and normal plasma.

ACID-BASE BALANCE

The normal metabolic processes of the body result in the continuous production of acids, such as carbonic, sulfuric, phosphoric, lactic, and pyruvic. In cellular oxidations the main acid end product is H_2CO_3 with 10 to 20 moles formed per day, which is equivalent to one to two liters of concentrated HCl. Although some foods yield alkaline end products, the acid type predominates, and the body is faced with the necessity of continually removing the large quantities of acids that are formed within the cells. An added restriction is that these products must be transported to the organs of excretion via the extracellular fluids without a great change in their H^+ concentration. The 7.35 to 7.45 pH range of blood is one of the most rigidly controlled features of the electrolyte structure. The means of accomplishing this is the mechanism of the regulation of acid-base balance, which involves water and electrolyte balance, hemoglobin and blood buffers, and the action of the lungs and the kidneys.

BODY BUFFER SYSTEMS

The ability of extracellular fluids to transport acids from the site of their formation in the cells to the site of their excretion in the lungs and kidneys without an appreciable change in pH depends on the presence of effective buffer systems in these fluids and in the red blood cells. A **buffer** is defined as a mixture of a weak acid and its salt that resists changes in pH when small amounts of acid or base are added to the system. The buffers in the plasma and extracellular fluid include the bicarbonate, phosphate, and plasma protein systems, which are represented as follows:

$$\frac{H_2CO_3}{BHCO_3} \qquad \frac{BH_2PO_4}{B_2HPO_4} \qquad \frac{H \text{ Plasma Protein}}{B \text{ Plasma Protein}}$$

The B stands for the base or cation.

In the red blood cells both bicarbonate and phosphate buffers are present along with two important hemoglobin buffers, as shown below:

$$\frac{H \text{ Hemoglobin}}{B \text{ Hemoglobin}} \qquad \frac{H \text{ Oxyhemoglobin}}{B \text{ Oxyhemoglobin}}$$

The buffers that are most effective in the regulation of acid-base balance are the bicarbonate, the plasma protein, and the hemoglobin buffers. Later in the chapter the important role of hemoglobin and its derivatives in the transportation of CO_2 from the tissues to the lungs without a change in pH will be considered.

The Bicarbonate Buffer

The bicarbonate buffer is by far the most important single buffer in acid-base balance. It is closely related to the constant production of CO_2, H_2CO_3, and $BHCO_3$; to the reactions of hemoglobin and oxyhemoglobin in the red cells; to the respiratory control of the H_2CO_3 concentration; and to the effect of the kidneys on $BHCO_3$ concentration. To illustrate the action of the buffer, reactions for the addition of a small amount of a strong acid and a strong base may be written as follows:

$$HCl + NaHCO_3 \rightarrow NaCl + H_2CO_3 \tag{1}$$
$$NaOH + H_2CO_3 \rightarrow H_2O + NaHCO_3 \tag{2}$$

In reaction (1), the basic member of the buffer pair reacts with the acid to form neutral NaCl and the acid member of the buffer pair. In reaction (2), the acid partner buffers the base to form water and the basic partner.

The buffering capacity of a buffer pair is related to its effectiveness in limiting changes of pH when acid or alkali is added to the system. The **Henderson-Hasselbalch equation** expresses the relation between the pH of the system, the pK of the acidic form of the buffer pair, and the concentration of each member of the buffer pair. The equation for the bicarbonate buffer is as follows:

$$pH = pK + \log \frac{[BHCO_3]}{[H_2CO_3]}$$

The pK is different for each buffer and is determined by measuring the pH of a solution that contains equal concentrations of the two components that make up the buffer pair. For example, if equal concentrations of $NaHCO_3$ and H_2CO_3 are present in a solution the pH equals 6.1, or

$$pH = pK = 6.1 \text{ when } \frac{[BHCO_3]}{[H_2CO_3]} = \frac{1}{1}, \text{ since } \log 1 = 0$$

Under normal conditions at the pH of blood, 7.4, the equation for the bicarbonate buffer is

$$7.4 = 6.1 + \log \frac{[BHCO_3]}{[H_2CO_3]} = 6.1 + \log \frac{27 \text{ mEq/l}}{1.35 \text{ mEq/l}} = 6.1 + \log \frac{20}{1} = 6.1 + 1.3$$

This unequal concentration of the buffer pairs would seem to indicate that the bicarbonate buffer is very ineffective at the pH of the blood. Because of the respiratory control of the H_2CO_3 concentration, however, buffering is remarkably effective.

The 1:20 ratio for H_2CO_3 and $BHCO_3$ in the plasma at pH 7.4 includes the normal values of 1.35 mEq/l for H_2CO_3 and 27 mEq/l for $BHCO_3$. In disease conditions, this ratio may be changed by increases or decreases in H_2CO_3 or $BHCO_3$, producing an acidosis or alkalosis. The most common changes involve the $BHCO_3$ concentration and are described as **metabolic acidosis** or **alkalosis.** Diseases that alter respiratory function affect the H_2CO_3 concentration and produce what is called **respiratory acidosis** or **alkalosis.** These are the four major abnormalities in acid-base balance.

HEMOGLOBIN

Hemoglobin is the respiratory protein of the red blood cell that has been described in detail by Perutz and his coworkers. This protein has a molecular weight of about 64,500, and is composed of four polypeptide chains and four heme molecules. Heme is a protoporphyrin derivative, with one iron atom coordinated with each of the four pyrrole nitrogen atoms. The four polypeptide chains, two α and two β chains, exist in the form of α helices which are folded and bent into three-dimensional structures. In the hemoglobin molecule, the four chains are grouped in a tetrahedron-shaped structure with the four heme molecules embedded in hollows in the folded chains. The relationship of the heme molecules and the polypeptide chains may be seen in the model of hemoglobin (Fig. 14–4).

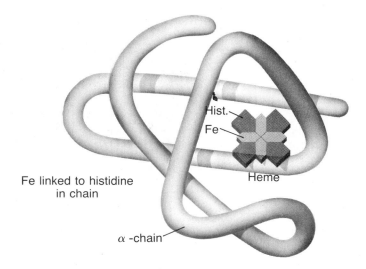

FIGURE 14-4 Artist's conception of the α-chain of the hemoglobin molecule with its associated heme group, illustrating the linkage of iron to histidine in the chain. (After Steiner: The Chemical Foundations of Molecular Biology. Princeton, N. J., D. Van Nostrand, 1965.)

The normal concentration of hemoglobin in the blood varies from 14 to 16g per 100 ml. This means that a 150-pound person would have a total of approximately 900g of hemoglobin. Since the red blood cells that contain the pigment are constantly being broken down, there is a continuous degradation of hemoglobin into other pigments in the body, such as **bilirubin,** which is converted into pigments responsible for the characteristic color of bile, urine, and feces.

Heme

TOPIC OF CURRENT INTEREST

THE ABNORMAL FORMS OF HEMOGLOBIN

In the last chapter we discussed three genetic disorders of amino acid metabolism, phenylketonuria, alkaptonuria, and tyrosinemia. Genetic variations originate through mutations, which are changes in the gene that are passed on to succeeding generations. Interest in the biochemical mechanisms of mutation began with the discovery in 1927 that irradiation induced mutations in cells. The many inborn errors of metabolism that have been described (p. 167) all show a genetic defect in enzymes responsible for normal metabolic reactions. One of the most striking examples of human genetic mutations is the various mutations of the hemoglobin molecule. Studies of abnormal forms of hemoglobin are facilitated by the ease of obtaining samples, and by the technique of peptide mapping of fragments from individual polypeptide chains. This technique is based on the two-dimensional paper chromatography described in Chapter 2. About 150 different kinds of mutant hemoglobins have been described in human blood. It is interesting to note that it has been concluded from studies on large groups of people that about 5 of every 1000 persons carry a gene for a mutant form of hemoglobin. Almost all of the genetic changes that have been observed in mutant hemoglobins are due to the replacement of a single amino acid, usually in either the α or the β chain. By comparing the genetic code or codon for the original amino acid in the chain and for the amino acid that replaces it, it has been determined that the change takes place in only one of the three bases in the nucleotide sequence in the codon. Table 14–1 lists some of the many abnormal hemoglobins and shows the position of the amino acid change, the chain in which the change occurs, and the change in codons.

In many instances, the amino acid replacement produces a chemical situation which causes the clinical symptoms in patients whose blood contains the abnormal hemoglobin. For example, in hemoglobin M Boston, the iron atom bound to the heme molecule is more stable in the ferric than in the ferrous state. Apparently, when tyrosine replaces histidine at position 58 in the α chain, it binds ferric iron more strongly than ferrous iron and stabilizes the ferric state. This prevents the ready reversibility of the formation of methemoglobin (ferric state); clinical symptoms are associated with the fact that methemoglobin does not carry oxygen.

Hemoglobin S has been studied in greater detail than the other abnormal forms. When glutamic acid in position 6 in the β chain is replaced by valine, this places two valines close to each other in positions 1 and 6. Apparently, a hydrophobic association results, which causes the hemoglobin S molecule to assume a conformation that distorts the shape of the erythrocyte itself. At low oxygen tension the erythrocytes assume a crescent shape instead of the flat disk shape of normal cells and are said to sickle, giving rise to the term for the human disease **sickle-cell anemia.** Most of the people with genes for hemoglobin S in their erythrocytes are **heterozygotes,** who carry one gene for normal hemoglobin and one gene for hemoglobin S; these people are said to possess the sickle-cell trait. Individuals who have two genes for hemoglobin S are known as **homozygotes** and suffer from sickle-cell anemia. Among Americans of

TABLE 14–1   AMINO ACIDS REPLACEMENTS IN SOME ABNORMAL HEMOGLOBINS

HEMOGLOBIN	POSITION IN CHAIN	AMINO ACID CHANGE		CODON CHANGE	
		From	*To*	*From*	*To*
		α Chain			
J Toronto	5	Ala	Asp	GCU	GAU
I	16	Lys	Glu	AAA	GAA
Memphis	23	Glu	Gln	GAA	CAA
Norfolk	57	Gly	Asp	GGU	GAU
M Boston	58	His	Tyr	CAU	UAU
		β Chain			
C	6	Glu	Lys	GAA	AAA
S	6	Glu	Val	GAA	GUA
G Saskatoon	22	Glu	Ala	GAA	GCA
M Emory	63	His	Tyr	CAU	UAU
Kansas	102	Asn	Thr	AAU	ACU

African descent, 1 of every 10 is a heterozygote, carrying one gene for hemoglobin S, whereas only 1 of 400 is homozygous for hemoglobin S. The crescent-shaped or sickled cells are viable only about half as long as normal cells and they clump together, especially at low oxygen tension. Clumps of these cells can block capillaries and decrease the blood supply to vital regions in the body. The majority of homozygotes die of sickle-cell anemia before the age of thirty, while the heterozygotes lead fairly normal lives. It is of interest from the standpoint of heredity and genetics that individuals with the sickle-cell trait are resistant to invasion by malaria parasites, an advantage in Africa where malaria is still common.

RESPIRATION

Hemoglobin is often called the **respiratory pigment** of the blood and has the property of combining with gases such as oxygen and carbon dioxide. The transportation of oxygen by the blood depends on the reversible reaction between hemoglobin and oxygen to form **oxyhemoglobin.**

$$Hb + O_2 \rightleftharpoons HbO_2$$
$$\text{Hemoglobin} \qquad \text{Oxyhemoglobin}$$

The oxygen capacity of the blood, about 1000 ml, is sufficient for normal tissue requirements. Some conception of the role of hemoglobin may be gained by a comparison of the oxygen capacity of plasma and whole blood. One liter of plasma can carry only 3 ml of oxygen in solution. In the absence of hemoglobin, the body's circulatory system would have to contain over 300 liters of fluid to supply oxygen to the tissues. This would represent a system four to five times our body weight.

In the process of respiration, hemoglobin comes into contact with a relatively rich oxygen atmosphere (partial pressure of 100 torr) in the alveoli of the lungs to form oxyhemoglobin. The oxyhemoglobin is carried by the arterial circulation to the tissues where a low oxygen concentration (partial pressure of 40 torr) and a high carbon dioxide concentration (partial pressure of 60 torr) combine to release the oxygen to the tissues. The carbon dioxide is then carried back to the lungs for excretion and the cycle is repeated.

URINE

In the previous chapters on metabolism several mechanisms that operate to maintain the constituents of the blood within fairly narrow limits of concentration have been considered. The utilization and transformation of nutrient material in the blood was emphasized in the chapters on metabolism. The removal of the waste products of metabolism, such as drugs, toxic substances, excess water, inorganic salts, and excess acid or basic substances, is essential to maintain the normal composition of the blood. The kidneys play a major role in the regular excretion of these substances from the blood and tissue fluids. The kidneys, therefore, are essential for the maintenance of blood and tissue fluid volume, the electrolyte and acid-base balance of the body, and the maintenance of normal osmotic pressure relationships of the blood and body fluids.

Water, carbon dioxide, and other volatile substances are eliminated from the body by the lungs. The skin excretes small amounts of water, inorganic salts, nitrogenous material, and lipids. Some inorganic salts are eliminated by the intestine, and the liver is involved in the excretion of cholesterol, bile salts, and bile pigments. Compared to the kidney, the other organs of excretion play a minor role.

The Formation of Urine

The kidney may be regarded as a filter through which the waste products of metabolism are passed to remove them from the blood. The blood enters the kidney by means of the renal arteries, which break up into smaller branches leading to the small filtration units called **malpighian corpuscles.** Each human kidney contains approximately 1,000,000 of these units. A malpighian corpuscle consists of a mass of capillaries from the renal artery which form the **glomerulus.** The glomerulus is enclosed within a capsule called **Bowman's capsule** which opens into a long tubule. Several of these tubules are connected to larger **collecting tubules** which carry the urine to the bladder. These anatomical structures of the kidney are illustrated in Figure 14–5.

The most generally accepted theory for the formation of urine can be outlined as follows: As the blood passes through the glomerulus, the constituents other than protein filter through the capillary walls and enter the tubules. As this filtrate passes down the tubules, a large proportion of the water and any substances which are of value to the body, such as glucose, certain inorganic salts, and amino acids, are reabsorbed into the blood stream. These substances are called **threshold substances.** Waste products of metabolism, such as urea and uric acid, are not completely reabsorbed, and therefore pass into the collecting tubules for excretion. A few substances, such as creatinine and potassium, are partially removed from the blood through excretion by the tubules in addition to filtration by the glomerulus. The function of the renal glomeruli may be regarded as that of ultrafiltration producing a protein-free filtrate of the plasma, followed by a process

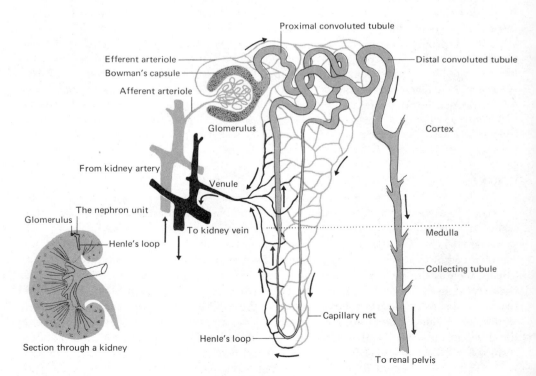

Figure 14–5   Diagram of a single kidney tubule and its blood vessels. (After Villee: Biology. 6th ed. Philadelphia, W. B. Saunders Company, 1972.)

of selective reabsorption by the renal tubules. Some concept of the magnitude of this process can be gained from the daily values for filtration and reabsorption. In a normal individual, 170 to 180 liters of water, 1000g of NaCl, 360g of $NaHCO_3$, and 170g of glucose are filtered through the glomeruli, and 168.5 to 178.5 liters of water, 988g of NaCl, 360g of $NaHCO_3$, and 170g of glucose are reabsorbed by the tubules in order to excrete about 30g of urea, 12g of NaCl, and other waste products in about 1500 ml of urine.

REGULATORY POWER OF THE KIDNEY

Although the kidney is considered mainly as an organ of excretion, it plays a regulatory role in water, electrolyte, and acid-base balance.

Water Balance

The control of the water content of blood, interstitial fluid, cell fluid, and digestive fluids in the body depends on normally functioning kidneys. Water in excess of body requirements is readily excreted by the kidney. When, however, water is needed to maintain the normal concentration of body fluids, the hypothalamus, in conjunction with the pituitary gland, secretes an antidiuretic hormone called **vasopressin.** This hormone causes an antidiuresis that results in less water being excreted in the urine and more being reabsorbed into the body fluids. Apparently changes in the osmotic pressure of the plasma regulate the secretion of vasopressin, which in effect is the "fine control" of urine volume.

Electrolyte Balance

To a certain extent the excretion or retention of electrolytes such as Na^+, K^+, Cl^-, HCO_3^-, and HPO_4^{-2} depends on the water balance of the body fluids. Electrolytes are excreted or reabsorbed along with the movement of water in the tubules. The steroid hormones of the adrenal cortex, such as **aldosterone** which is called a **mineralocorticoid,** exert more specific control over the excretion of electrolytes by the tubules. When the plasma has too much water, its osmotic pressure decreases, and it is said to be hypotonic. Under these conditions there is an increased secretion of adrenal cortex hormones that increase the retention or reabsorption of electrolytes into the plasma. If the plasma electrolytes were too concentrated and the osmotic pressure increased, the secretion of the cortical hormones would be depressed and the secretion of vasopressin would increase to assist in the excretion of electrolytes and the retention of water.

Acid-Base Balance

In the tubules Na reabsorption, in part at least, is involved in the regulation of acid-base balance through the action of the Na^+—H^+ exchange. This phase of Na^+ excretion is regulated to some extent by the pH of the blood plasma and the capacity of the tubular cells to acidify the urine by H^+ and NH_3 formation.

The glomerular filtrate contains the electrolytes and acids and bases present in the plasma. The chief cation is Na^+, and the chief anions are Cl^-, HCO_3^-, and HPO_4^{-2}, as electrolytes of NaCl, $NaHCO_3$, and Na_2HPO_4. At a pH of 7.4, about 95 per cent of the CO_2 is present as $NaHCO_3$ and about 83 per cent of the PO_4^{-3} as Na_2HPO_4. Normally, as the glomerular filtrate passes into the tubules most of the water is reabsorbed, and the greater proportion of the Na^+ is taken up by the tubular cells in exchange for H^+ formed in the tubular cells from H_2CO_3. This Na^+ is returned to the plasma in association with the HCO_3^- formed from H_2CO_3 in the tubular cells. The formation of H^+ and HCO_3^- from H_2CO_3 in both the proximal and distal tubular cells is catalyzed by the enzyme carbonic anhydrase, as follows:

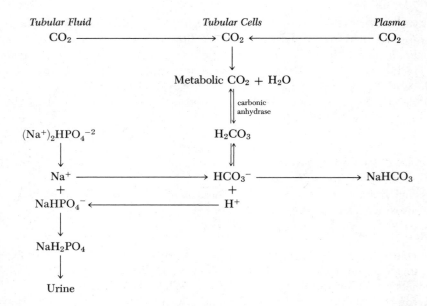

By varying the rate of ventilation, the lungs control the H_2CO_3 concentration in the plasma.

The task of the kidneys is to stabilize the concentration of the bicarbonate. This problem involves two aspects: first, the salvaging of all, or nearly all, of the bicarbonate contained in the glomerular filtrate (equivalent to approximately one pound of $NaHCO_3$ per 24 hours); and second, the neutralization of nonvolatile acids (H_2SO_4 and H_3PO_4). The kidney may conserve base in two ways: by conversion of neutral or basic salts to acid salts for excretion, as shown previously for Na_2HPO_4, and by the synthesis of ammonia.

The mechanism for the synthesis of ammonia to further conserve Na^+ and fixed base may be represented as follows:

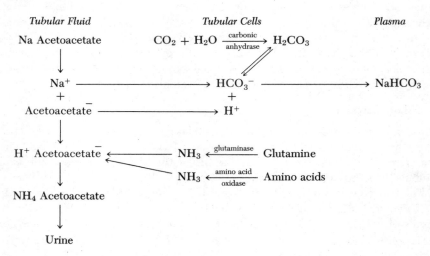

The urinary excretion of the ammonium ion in a person on a normal average diet is 30 to 70 mEq per day. In conditions of acidosis that might result from uncontrolled diabetes mellitus, acetone bodies, such as acetoacetic acid, are formed. The sodium salt of this acid must be excreted by the kidney, and to conserve the Na^+ for plasma buffers H^+ and NH_3 are formed in the tubular cells of the kidney.

IMPORTANT TERMS AND CONCEPTS

acid-base balance
bicarbonate buffer
blood
electrolyte balance
electrolytes
extracellular fluids
hemoglobin

plasma
plasma proteins
regulatory power of kidney
respiration
sickle-cell anemia
urine formation
water balance

QUESTIONS

1. List the important functional body fluids and their volumes in a normal adult.

2. What is oncotic pressure, and how does it assist in the normal distribution of fluids between the circulatory system and the interstitial space?

3. Outline the parameters involved in the balance between fluid intake and fluid output in the body.

4. Explain the difference between whole blood, plasma, and serum.

5. Compare an electrophoretic pattern of plasma proteins with an immunoelectrophoretic pattern. What information may be gained from each pattern?

6. Illustrate with a bar graph the electrolyte composition of normal plasma.

7. What are the most important buffers in the whole blood and plasma?

8. Calculate the pH of a bicarbonate buffer solution that contains 20mEq/l of $BHCO_3$ and 2mEq/l of H_2CO_3.

9. What is the chemical nature of heme, and what is its relation to hemoglobin?

10. Briefly discuss the genetic changes responsible for the formation of abnormal hemoglobins.

11. List the essential functions of the kidney and describe the process of formation of the urine.

12. What hormones are involved in the control of water and electrolyte balance by the kidney? Explain how they function.

13. Explain how the process of H^+ formation in the kidney tubules helps conserve Na^+ for the plasma.

14. Why would the kidney tubules increase the synthesis of ammonia in the condition of diabetes mellitus? Explain.

Chapter 15
BIOCHEMISTRY
OF DRUGS

The *objectives* of this chapter are to enable the student to:

1. Explain the use of aspirin in so many drug preparations.
2. Recognize the similarities and differences between the many drugs used as tranquilizers or sedatives.
3. Explain the increased use of amphetamine-type drugs in today's society.
4. Recognize the difference in action and potency of marijuana, LSD, and heroin.
5. Discuss the therapeutic uses of the antihistamines.
6. Compare the action of antibacterial and antibiotic drugs.
7. Describe the mechanism of action of the antifertility drugs.
8. Explain the use of cytotoxic drugs in the treatment of cancer.

If we accept a single definition of disease as *dis ease,* headaches, minor aches and pains, malnutrition, dietary deficiencies, metabolic abnormalities, endocrine disturbances, infections, and cancer all qualify as diseases. Volumes have been written on diseases of the cell, tissues and organs of the body, and the therapeutic agents or drugs used to combat the disease process. It is becoming more and more apparent that all disease has a biochemical basis. The biochemistry of normal and abnormal heredity, deficiency diseases, errors of metabolism, and the process of infection is receiving considerable attention, study, and research.

ANALGESIC DRUGS

To aid in the understanding of the recent emphasis on health-related research, examples of different types of diseases and their treatment will be discussed. So many people are occasionally inconvenienced with headaches and minor aches and pains that their cause and treatment with analgesic drugs should be of interest. In general, these pains are caused by swelling of tissue, resulting in pressure on peripheral nerves, and also minor inflammation of tissues, accompanied by an increase in body temperature which affects nerve endings. A common analgesic drug that serves as the basis for a multitude of headache, cold, flu remedies is acetylsalicylic acid or **aspirin.** When combined with **phenacetin** and **caffeine,** the resultant preparation is the common APC tablets. The compound **N-acetyl-p-amino phenol** (acetaminophen) is a metabolic product of phenacetin and has replaced this drug in several preparations. Phenacetin is an organic amine derived from acetanilide, a compound originally used as an antipyretic drug. Aspirin is an

186

analgesic drug in that it reduces inflammation and swelling of tissues, exerts an antipyretic action in reducing fever, and probably exerts a chemical action on the peripheral nerves. Phenacetin and N-acetyl-p-aminophenol exhibit some of the effects of aspirin and are synergistic with respect to its action. The antipyretic effect of aspirin is thought to be related to the salicylic acid portion of the structure, whereas the antipyresis exhibited by phenacetin and N-acetyl-p-aminophenol depends on the amino-benzene portion of the

| Aspirin | Phenacetin | N-acetyl-p aminophenol | Caffeine |

compound. Caffeine is a diuretic and assists the kidney in excretion of the drug and the circulatory system in the transport of the drug. The addition of buffering agents to aspirin has been found to speed the absorption of the drug into the blood and tissues and has resulted in products such as Bufferin.

TRANQUILIZERS, SEDATIVES, AND HALLUCINOGENIC DRUGS

Diseases of the brain and nervous system ranging from undue concern and nervousness through inability to sleep, anxiety symptoms, neurotic behavior, and pathology of brain tissue are common conditions in these accelerated times. Drugs such as sedatives, tranquilizers, and psychic energizers are too commonly prescribed and used. Mind-expanding and hallucinogenic drugs are receiving considerable attention, as we know. Another drug, L-dopa, provides substantial relief in Parkinson's disease, a condition involving changes in the brain tissue.

Phenothiazine, which has a three-ringed structure in which two benzene rings are linked by a sulfur atom and a nitrogen atom, forms the basis for a group of drugs that are potent adrenergic blockers. The phenothiazines are widely used in the treatment of pyschiatric patients and in the treatment of nausea and vomiting. There is a close relationship between the chemical structure and the activity of the drug. Substitution of a chlorine or methoxy group in position R′ in the basic structure increases the potency of the drug for depressing motor activity and altering psychotic behavior in patients. A CF_3 substitution in this position increases antiemetic and antipsychotic potency. One of the most potent phenothiazines has a CF_3 on position R′ and a piperazine group on position R″. Chlorpromazine is the most frequently prescribed phenothiazine, while fluphenazine is one of the most potent members of this group. Their relationship to the basic structure is shown as follows:

Phenothiazine
(Basic structure)

Chlorpromazine

Fluphenazine

Chlorpromazine also has a sedative effect and reduces the blood pressure.

Other compounds used as mild tranquilizers or sedatives, such as barbiturates and meprobamate, are related to the pyrimidines and to urea. These drugs depress the central nervous system. In the case of barbiturates their speed of action and duration of effect depend on an increased lipid solubility and an increase in the length of the side chain, as typified by the structure of Seconal versus that of Barbital. The pharmacological effects of meprobamate are very similar to those of the barbiturates.

Barbital   Seconal   Meprobamate

Continued use of mild sedatives or tranquilizers, like barbiturates, leads to a dependency on the drug. Overdosage or combinations of alcoholic beverages and barbiturates may lead to coma and accidental death. The more potent tranquilizers are subjected to strict control by physicians and are not as widely prescribed. When drugs with strong analgesic and sedative properties are required, narcotic drugs, such as **morphine** and **Demerol**, are employed. Morphine, first isolated from the opium poppy, has a complex chemical structure containing a phenanthrene nucleus and a piperidine nucleus. **Codeine,** which is also commonly used as a weaker narcotic, is a methyl ether of morphine. **Heroin** is the diacetyl ester of morphine and is used by drug addicts because it is more lipid-soluble and faster-acting than morphine. Demerol was synthesized as a substitute for morphine but proved to be no less habit-forming.

Morphine   Demerol

Antidepressant or mood-elevating drugs, sometimes called psychic energizers, are being prescribed in increasing quantities. Derivatives of hydrazine directly affect brain and nervous system function by inhibiting the enzyme monoamine oxidase, which apparently results in an antidepressant action. Iproniazid and **isocarboxazid** are examples of potent drugs in this family. Amphetamine, also called benzedrine, and related drugs such as dextroamphetamine, methamphetamine, and ephedrine all exert powerful central nervous system-stimulating actions. These include a decreased sense of fatigue, increased initiative and ability to concentrate, often an elevation of mood, elation and euphoria, and increased motor activity. Such properties have resulted in the use of these drugs as psychic energizers by truck drivers on long trips and by students cramming for exams. Unfortunately, prolonged use of these drugs is almost always followed by mental depression and fatigue. The drugs are also used as appetite depressants in obesity and to counteract depressive syndromes and behavioral syndromes.

Isocarboxazid

Amphetamine

A long step further in the use of mind-influencing drugs is represented by the **psychedelic** or **hallucinogenic drugs.** It is difficult to characterize the medical effect of these drugs, although many nonmedical experiments are being conducted. In view of the change in personality and mental state of an individual taking this type of drug, the results of ingestion are often unpredictable. **Marijuana** represents a mild type of hallucinogen and is obtained from the *Cannabis sativa* or hemp plant. The most potent marijuana is obtained from the yellow resin produced from the flowers of the ripe plant and is called **hashish.** Chemically the drug is a derivative of an alcohol, **cannabinol,** and the active constituent is believed to be a delta-L form, which has recently been synthesized. **Mescaline** and **lysergic acid diethylamide (LSD)** are examples of more potent hallucinogens.

Cannabinol

Mescaline

Lysergic acid diethylamide (LSD)

For years, there had been little advance in the treatment of **Parkinson's disease.** This is a disease of the brain affecting the metabolism of dopamine, epinephrine, and norepinephrine; it is also known as paralysis agitans, since it involves both a progressive paralytic rigidity and tremors of the extremities. Recently it was found that the dopamine concentration of certain areas of the brain was markedly deficient in chronic patients who had died of the disease. Since dopamine will not penetrate the brain tissue when carried in the blood, the precursor L-dihydroxyphenylalanine, L-dopa, was administered and found to increase the dopamine concentration in the target areas. The use of the drug L-**dopa** provides considerable relief and improvement of symptoms for these patients and has been hailed as a breakthrough in Parkinson's disease therapy.

L-Dopa
(L-dihydroxyphenylalanine)

Dopamine

ANTIHISTAMINES

Histamine, which is formed by the decarboxylation of the amino acid histidine (see p. 12), is a powerful pharmacological agent with effects on the vascular system, smooth muscles, and exocrine glands, especially the gastric glands. The administration or release of histamine causes dilatation of capillaries and small blood vessels with a subsequent drop in systemic blood pressure; the dilatation of cerebral vessels results in a histamine headache which may be very severe. Smooth muscles, especially the bronchioles, are stimulated by histamine and may cause respiratory problems in persons suffering from bronchial asthma and other pulmonary diseases. Histamine is a powerful gastric secreto-gogue and produces a copious secretion of gastric juice of high acidity. It also stimulates nerve endings and causes itching when introduced into the superficial layers of the skin.

Antihistamines are drugs that antagonize the pharmacological actions of histamine and also reduce the intensity of allergic reactions. A portion of the chemical structure of various antihistamines is similar to that in histamine, and these drugs act as competitive antagonists to histamine. Apparently they occupy the receptor sites on the effector cells and exclude histamine from these sites. The common core of the chemical structure in both histamine and antihistamines is a substituted amine, i.e., ethylamine. It is believed that it is this portion of the molecule that competes with histamine for the receptors.

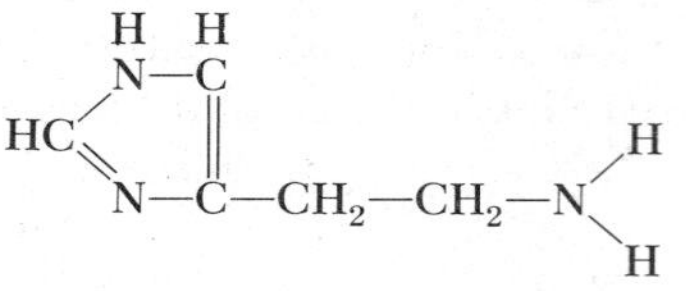

Histamine

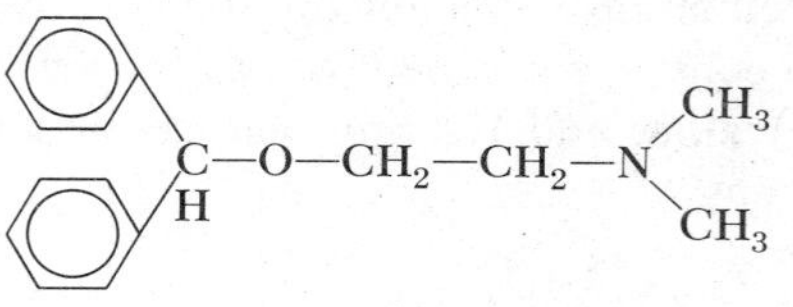

Benadryl (Diphenhydramine)

Therapeutically the antihistamines are most commonly used in the symptomatic treatment of various allergic diseases. Patients with bronchial asthma, hay fever, allergic rhinitis, and chronic rhinitis with superimposed acute colds gain considerable relief by the use of antihistamines. Various types of allergic dermatitis, contact dermatitis, insect bites, and poison ivy are benefited by the topical application of antihistamine-containing lotions. Some of these drugs, especially **Dramamine,** which relies on diphenhydramine as the active agent, are very effective against motion sickness. One common side effect of most antihistamines is their tendency to induce sedation, which restricts their daytime use when it is necessary to operate motor vehicles. **Chlor-trimeton,** shown below, is less prone to produce drowsiness than most other preparations. The prominent hypnotic effect of antihistamines related to **Benadryl** and **Pyribenzamine** has resulted in their use in various proprietary remedies for insomnia, such as Sominex and Nytol.

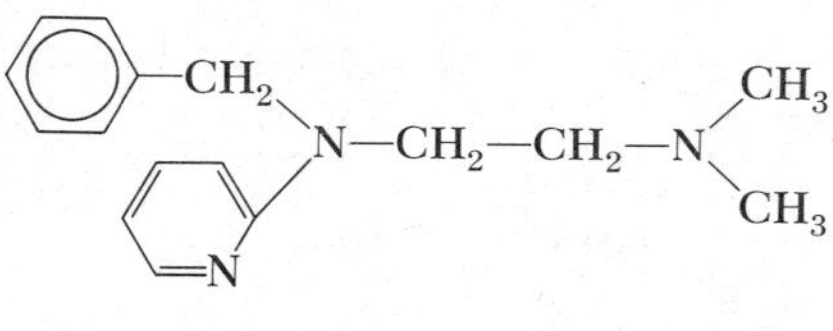

Pyribenzamine (Tripelennamine)

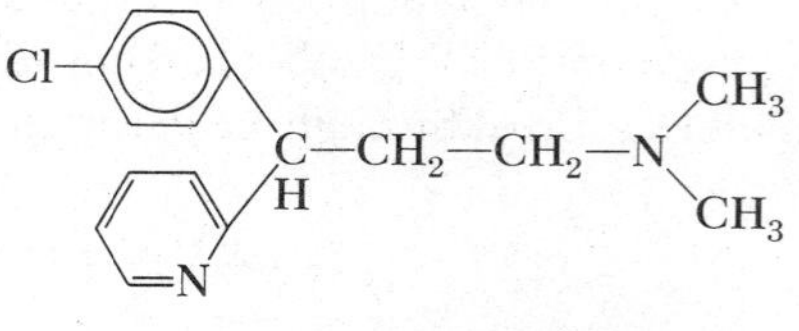

Chlor-trimeton (Chlorpheniramine)

ANTIBACTERIAL AND ANTIBIOTIC DRUGS

The body has to guard against infection by bacteria or viruses from birth to death. The newborn infant is fortified with antibodies against disease upon receiving the gamma globulins in the mother's milk. The layer of skin covering the body, the hydrochloric acid of the gastric juice, the digestive enzymes, and the various phagocytic cells in the circulation, all serve as a first-line defense against infection. Vaccination during childhood stimulates the production of antibodies to certain diseases. Until the mid-1930s, a serious infection or infectious disease was viewed with alarm by physicians and laymen alike. The first major group of chemotherapeutic agents were the sulfonamides or **sulfa drugs,** prepared by the acylation of sulfanilamide. These drugs were antibacterial agents that inhibited the synthesis of a compound like folic acid that was essential for the continued growth of the invading bacteria. Sulfanilamide, for example, acts as a competitive inhibitor of the enzyme that is involved in the utilization of *p*-aminobenzoic acid in the synthesis of tetrahydrofolic acid by the bacteria (see p. 51). **Sulfanilamide** was found to be effective in the treatment of streptococcus infections, pneumonia, puerperal fever, gonorrhea, and gas

| Sulfanilamide | Sulfaguanidine | Sulfathiazole | Sulfadiazine |

gangrene. The drug is only slightly soluble in water and may damage the kidney by accumulation in that organ during excretion. Other toxic reactions, including methemoglobinemia, resulted in the development of other derivatives, such as **sulfaguanidine, sulfathiazole,** and **sulfadiazine.** A thorough study of the therapeutic properties of each sulfa drug resulted in better treatment and control of various infectious diseases. Sulfadiazine, for example, is less toxic than the other sulfa drugs, yet is one of the most effective in the treatment of pneumonia and staphylococcus infections.

A few years after the development of the sulfa drugs, a new type of antimicrobial agent was accidently discovered by Fleming. He observed that a staphylococcus culture on a bacterial plate did not grow around the periphery of a blue-green mold that had contaminated the culture plate. **Penicillin** was isolated from the secretion of the mold and was termed an antibiotic, since it interfered with the growth of the bacteria. A large number of antibiotic agents have been isolated from similar experiments with other molds. It required nine years of intensive research to synthesize penicillin. Other antibiotics, including **streptomycin, tetracycline,** and **prostaphlin,** have been synthesized, and prostaphlin shows considerable promise as an effective control of infections caused by staphylococcal bacteria. As in the case of the sulfa drugs, a family of antibiotics with specific antimicrobial properties has now been developed.

| Penicillin G | Tetracycline |

STEROID DRUGS IN RHEUMATOID ARTHRITIS

There are several types of collagen diseases associated with inflammatory changes in connective tissue which affect mainly the joints, skin, heart, and muscle. **Rheumatoid arthritis** and **lupus erythematosus** are examples of collagen disease. Arthritis is most common and affects men and especially women in their forties and fifties. For many years, aspirin has been used as a mild antiinflammatory agent in arthritis and continues as the maintenance therapy. More recently **cortisone** has been found to reduce inflammation of the joints in arthritis and to reverse the course of this and other collagen diseases. The dosage level required to produce these desirable effects also produced several unwanted side effects. Steroid derivatives of cortisone were developed to decrease the incidence of side effects and increase the therapeutic potency of the drugs. The 9-fluoro-16-methyl derivative of **prednisolone,** a steroid closely related to cortisol, is 100 to 250 times as potent as cortisone in the treatment of rheumatoid arthritis. At present, arthritic patients are maintained with aspirin and given small doses of potent steroid drugs whenever an acute inflammatory process flares up in their joints. The relationship between the structure and the function of cortisone and related steroids has already been discussed in Chapter 6 (see p. 85).

Prednisolone   9-Fluoro-16-methyl prednisolone

ANTIFERTILITY DRUGS

Another type of steroid drug related to the sex hormones is the **antifertility drugs.** These drugs are unique in that they are given to inhibit a normal physiological process, whereas the great majority of drugs are used to treat a disease process or to alleviate the symptoms of a disease. As early as 1937 it was shown that the hormone progesterone would inhibit ovulation in rabbits. In the 1950s, several laboratories attempted to develop orally active steroids that possessed the properties of progesterone. **Norethindrone, 17-α-ethynyl-19-nortestosterone,** and **norethynodrel,** the progestational component of Enovid, resulted from these studies. These drugs were related to testosterone (see p. 86) and contained a 17-α-ethynyl group.

Norethindrone   Norethynodrel

Since progesterone is relatively inactive when given orally, derivatives of this compound were studied for oral potency. It was found that 17-acetoxy-progesterone was active orally and that the addition of an α-methyl group to produce **medroxyprogesterone acetate** further enhanced this activity.

In clinical trials in Puerto Rico, Haiti, and the United States, the testosterone and progesterone derivatives described previously were found to effectively suppress ovulation in women. Also, it was discovered that estrogen enhanced the suppressive effect of the progesterone and that the 3-methyl ether of ethynyl estradiol or **mestranol** served as a potent estrogen. A combination of orally effective progesterone and estrogen active drugs is therefore commonly used in "the pill." The original dose was 10 mg of the progesterone

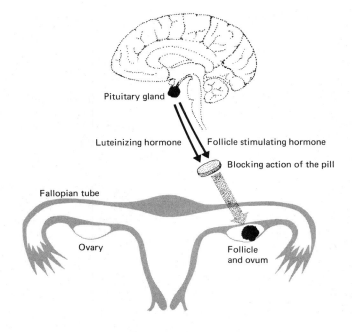

Figure 15-1   The effect of oral contraceptives on the process of ovulation.

Mestranol

Medroxyprogesterone acetate

and 0.15 mg mestranol, but this is being reduced toward 1 mg progesterone and 0.05 mg mestranol to decrease the occurrence of such side effects as nausea, headaches, dizziness, and thrombosis. The drug is usually taken on days five to 25 of the menstrual cycle and then withdrawn to permit normal menstruation.

From a study of the mode of action of oral contraceptives it was concluded that the primary effect is inhibition of follicular maturation (see p. 85), which prevents the occurrence of ovulation.

This mechanism is outlined in Figure 15-1, which illustrates the blocking action of the progesterone derivatives on the hormones of the pituitary gland responsible for development of an ovarian follicle and the subsequent release of an unfertilized ovum by the process of ovulation.

HYPOGLYCEMIC DRUGS IN DIABETES MELLITUS

Diabetes mellitus, because of its frequency, is probably the most important metabolic disease. The fundamental difficulty in the disease is a relative or complete lack of insulin, which is necessary for normal carbohydrate metabolism. Since the metabolic pathways of carbohydrates, fats, and proteins are known to be closely interwoven, any essential fault in carbohydrate metabolism also involves the metabolism of fat and protein, as well as water, electrolyte, and acid-base balance. There is evidence that diabetes is a hereditary disease and that the genetic tendency toward the disease results in the common condition known as prediabetes, which exists in many relatives of diabetics. The chemistry of **insulin** (p. 19) and its function in diabetes (p. 121) have already been discussed. The beta cells of the islet tissue in the pancreas produce insulin, and any condition resulting in hyperglycemia stimulates the pancreas to secrete greater quantities of this hormone. To control diabetes, insulin must be injected into the muscle tissue daily. In view of the discomfort and inconvenience to the patient, many efforts have been made to prolong the action of insulin and to develop drugs with insulin-like activity that can be taken orally. In 1942 it was observed that a sulfamidothiazole compound exhibited a potent hypoglycemic effect. Related compounds were tested, and by 1955 sulfonylurea derivatives such as **tolbutamide** (Orinase) were available as antidiabetic agents. Guanidine derivatives also produced hypoglycemia, and **phenethylbiguanide** (Phenformin) represents another type of antidiabetic drug. When administered orally these two drugs exhibit different properties.

Orinase (Tolbutamide)   Phenformin (Phenethylbiguanide)

Orinase stimulates the secretion of insulin from the beta cells, while Phenformin stimulates the oxidation of glucose by the peripheral tissues. Both of the drugs are effective in the treatment of diabetic patients over 40 years of age with stable and mild diabetes. Their use is ineffective in young unstable diabetics, especially those prone to ketoacidosis. In patients that do not respond to Orinase alone, the concurrent administration of both drugs is often effective. It has also been found that diabetics with congestive heart failure or severe renal insufficiency should not be treated with either of these drugs.

URICOSURIC DRUGS IN GOUT

Gout is a chronic disease that is characterized by an increased level of uric acid in the blood (hyperuricemia) acute episodes of gouty arthritis, and degenerative changes in the joints. The disease is seen predominantly in men and the initial stages occur in their forties. In the early days, gout appeared to be more prevalent among the aristocracy, and at present, professors and physicians exhibit the greatest incidence of the disease; however, even vegetarians and lower economic groups are susceptible to the disease.

In the treatment of the disease an attempt is made to decrease the level of uric acid in the blood and in the body stores. **Probenecid** (Benemid), a derivative of benzoic acid, was found to inhibit the enzymes responsible for the reabsorption of uric acid by the kidney. It is a safe drug that lowers serum uric acid by about 50 per cent within

Benemid   Allopurinol

two to four days and maintains the reduced level as long as therapy is continued. A more recent therapeutic agent is **allopurinol** that inhibits the enzyme xanthine oxidase and thus reduces the formation of uric acid from its immediate precursors, hypoxanthine and xanthine. Treatment with allopurinol maintains the blood uric acid level at a normal value and lowers the body pool of uric acid by constant excretion through the kidney. Both of these drugs were first "tailor-made" by pharmacologists for a specific biochemical purpose. Probenecid was developed to decrease the renal tubular secretion of penicillin, and was found to be very effective in decreasing the reabsorption of uric acid by the tubules. Allopurinol was synthesized to serve as a potent inhibitor of xanthine oxidase to prevent the oxidation of 6-mercaptopurine and thus extend its effective action in the treatment of leukemia.

TOPIC OF CURRENT INTEREST

DRUGS IN THE TREATMENT OF CANCER

Cancer is a general term used to describe rapid multiplication of cells and increased growth of certain tissues in the body. Cancers are also called **tumors** and **neoplasms** and are classed as malignant when they spread to other parts of the body and exhibit recurrence of growth after surgery. **Leukemia** is a cancer of blood-forming tissues characterized by abnormal leucocytes or white blood cells; **carcinoma** involves epithelial cells; and **sarcoma** is a tumor of muscle or connective tissue. Considerable research, effort, and money have been expended in extensive studies of cancer in recent years. Tumors in experimental animals have been produced by **carcinogenic agents** such as dimethylbenzanthracene and have been transplanted in normal animals to aid in the study of the metabolism of cancer cells and the development of therapeutic agents. The presence of possible carcinogenic agents in food additives, public water supplies, and plastics is presently being investigated by environmental pollution specialists and toxicologists. Although surgical removal of tumors and irradiation with x-rays and radioisotopes constitute two major modes of attack on cancer, several classes of chemical compounds, including alkylating agents, anti-metabolites, and purine and pyrimidine analogs, show considerable promise in cancer treatment.

Nitrogen-containing compounds related to mustard gas (dichloroethyl sulfide) are alkylating agents that have been used in cancer therapy for several years. To overcome undesirable side effects of these drugs, **cyclophosphamide** was synthesized; it is a cytotoxic agent that has shown good initial results in the treatment of Hodgkin's disease and lymphosarcoma. When used with other drugs, cyclophosphamide has produced complete remissions of acute lymphoblastic leukemia in children for more than five years. An example of an antimetabolite used in therapy is **methotrexate**, a folic acid analog. This drug produces dramatic temporary remissions of leukemia in children and long-lasting remissions in choriocarcinoma (an epithelial tumor that occurs in the uterus at the placental site during pregnancy).

The purine analog **6–mercaptopurine** is believed to suppress the biosynthesis of purines within the cell, thus interfering with the production of RNA and DNA in the tumor cells. This drug, when used in combination with other drugs, is the most

6-Mercaptopurine   Cyclophosphamide

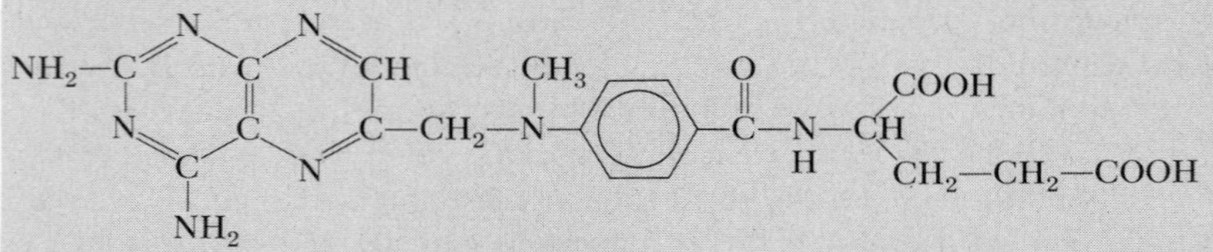

Methotrexate

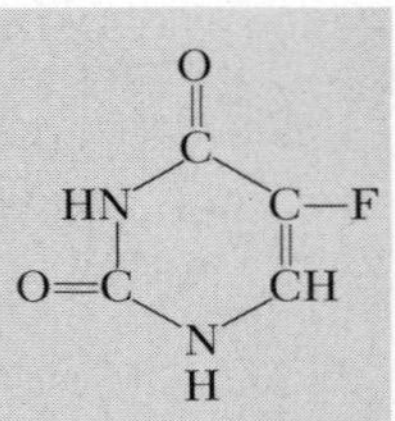

Fluorouracil

valuable purine analog for the treatment of acute leukemia. **Fluorouracil**, a pyrimidine analog, is particularly effective in the treatment of advanced carcinoma, especially of the breast and the gastrointestinal tract.

The structures of some of the cytotoxic drugs used in the treatment of cancer are shown above and on page 195.

Compounds extracted from natural products, such as **vinblastine** and **vincristine** from the periwinkle plant, are also used in cancer therapy. Vinblastine, a complex alkaloid, has been found to be effective in the treatment of choriocarcinoma in women. An interesting development in the chemotherapy of cancer is the discovery that the enzyme **L-asparaginase** is an effective agent in the treatment of leukemia.

At present it is difficult to assess the long-range benefits of chemotherapy in cancer. Millions of dollars and uncounted hours are being expended in the search for improved therapy, possible prevention, and ultimately, a cure for cancer.

IMPORTANT TERMS AND CONCEPTS

analgesic
antibiotics
antifertility drugs
antihistamine
cytotoxic drugs

heroin
hypoglycemic drugs
marijuana
sulfa drugs
tranquilizers

QUESTIONS

1. Why can it be stated that all disease has a biochemical basis? Explain.

2. Why is aspirin used in so many proprietary drug preparations? Explain.

3. Why are amphetamine-type drugs used in such large quantities today? Explain.

4. Isocarboxazid is an example of a monoamine oxidase inhibitor. Describe the usefulness of the drug.

5. Why are modifications of the phenothiazine structure important therapeutically? Explain.

6. Compare meprobamate and the barbiturates on the basis of their chemical structure, potency, and mode of action.

7. Give two examples of a hallucinogenic drug. Should the use of these drugs be controlled by law? Explain.

8. Discuss the chemical nature and action of L-dopa in Parkinson's disease.

9. Outline the main uses of antihistaminic drugs.

10. Explain the differences and similarities of sulfa drugs and antibiotics such as penicillin.

11. Give an example of a steroid drug used in the treatment of arthritis and explain its action.

12. Describe the mechanism of action of the antifertility drugs.

13. Explain the action of one oral insulin substitute and its advantages.

14. Compare the mechanism of action of Benemid and allopurinol in the treatment of gout.

15. Why is 6-mercaptopurine used in the treatment of leukemia?

16. Compare the chemical structure of allopurinol and 6-mercaptopurine. Should allopurinol be effective in the treatment of leukemia? Explain.

SELECTED READINGS

Chapter 1. Biochemistry of the Cell
Capaldi: A Dynamic Model of Cell Membranes. Scientific American, Vol. 230, No. 3, p. 26, 1974.
Everhart and Hayes: The Scanning Electron Microscope. Scientific American, Vol. 226, No. 1, p. 54, 1972.
Goodenough and Levine: The Genetic Activity of Mitochondria and Chloroplasts. Scientific American, Vol. 223, No. 5, p. 22, 1970.
Raff: Cell-Surface Immunology. Scientific American, Vol. 234, No. 5, p. 30, 1976.
Satir: The Final Steps in Secretion. Scientific American, Vol. 233, No. 4, p. 28, 1975.
Stent: Cellular Communication. Scientific American, Vol. 227, No. 3, p. 42, 1972.

Chapter 2. Proteins
Capra and Edmundson: The Antibody Combining Site. Scientific American, Vol. 236, No. 1, p. 50, 1977.
Delwiche: The Nitrogen Cycle. Scientific American, Vol. 223, No. 3, p. 136, 1970.
Edelman: The Structure and Function of Antibodies. Scientific American, Vol. 223, No. 2, p. 34, 1970.
Fraser: Keratins. Scientific American, Vol. 221, No. 2, p. 86, 1969.
Isenberg and Grdinic: Cyclic Disulfides, Their Function in Health and Disease. Journal of Chemical Education, Vol. 49, No. 6, p. 392, 1972.
McGuinness: Estimation of Protein Size, Weight, and Asymmetry by Gel Chromatography. Journal of Chemical Education, Vol. 50, No. 12, p. 826, 1973.
Research Reporter: Methionine and Origin of Life. Chemistry, Vol. 46, No. 2, p. 14, 1973.
Safrany: Nitrogen Fixation. Scientific American, Vol. 231, No. 4, p. 64, 1974.
Sharon: Glycoproteins. Scientific American, Vol. 230, No. 5, p. 78, 1974.
Vedvick and Coates: Hemoglobin: A Simple "Backbone" Type of Molecular Structure. Journal of Chemical Education, Vol. 48, p. 537, 1971.

Chapter 3. Nucleic Acids
Campbell: How Viruses Insert Their DNA into the DNA of the Host Cell. Scientific American, Vol. 235, No. 6, p. 102, 1976.
Crick: The Genetic Code. Scientific American, Vol. 215, No. 4, p. 55, 1966.
Research Reporter: Portarit of a Gene. Chemistry, Vol. 42, No. 8, p. 20, 1969.
Research Reporter: Further Confirmation of Nucleic Acid Double Helix. Chemistry, Vol. 46, No. 7, p. 19, 1973.
Research Reporter: New DNA Structure Proposed. Chemistry, Vol. 47, No. 9, p. 22, 1974.
Sobell: How Actinomycin Binds to DNA. Scientific American, Vol. 231, No. 2, p. 82, 1974
Temin: RNA-Directed DNA Synthesis. Scientific American, Vol. 226, No. 1, p. 24, 1972.
Watson: Double Helix. Atheneum, New York. 1968.
Yanofsky: Gene Structure and Protein Structure. Scientific American, Vol. 216, No. 5, p. 80, 1967.

Chapter 4. Enzymes
Koshland: Protein Shape and Biological Control. Scientific American, Vol. 229, No. 4, p. 52, 1973.
Miller and Cory: Activation Energies for a Base-Catalyzed and Enzyme-Catalyzed Reaction. Journal of Chemical Education, Vol. 48, p. 475, 1971.
Research Reporter: First Synthesis of an Enzyme, Ribonuclease. Chemistry, Vol. 42, No. 4, p. 21, 1969.
Stroud: A Family of Protein-Cutting Proteins. Scientific American, Vol. 231, No. 1, p. 74, 1974.
Thayer: Some Biological Aspects of Organometallic Chemistry. Journal of Chemical Education, Vol. 48, No. 12, p. 807, 1971.
Wroblewski: Enzymes in Medical Diagnosis. Scientific American, Vol. 205, No. 2, p. 99, 1961.

Chapter 5. Carbohydrates
Elias: The Natural Origin of Optically Active Compounds. Journal of Chemical Education, Vol. 49, No. 7, p. 448, 1972.
McCord and Getchell: Cotton. Chemical & Engineering News, November 14, 1960, p. 106.
Mowery: Criteria for Optical Activity in Organic Molecules. Journal of Chemical Education, Vol. 46, p. 269, 1969.
Slocum, Sugarman, and Tucker: The Two Faces of D and L Nomenclature. Journal of Chemical Education, Vol. 48, p. 597, 1971.
Vennos: Construction and Uses of an Inexpensive Polarimeter. Journal of Chemical Education, Vol. 46, p. 459, 1969.

Chapter 6. Lipids
Beyler: Some Recent Advances in the Field of Steroids. Journal of Chemical Education, Vol. 37, p. 497, 1960.
Gaucher: An Introduction to Chromatography. Journal of Chemical Education, Vol. 46, p. 729, 1969.
Karasek, DeDecker, and Tiernay: Qualitative and Quantitative Gas Chromatography for the Undergraduate. Journal of Chemical Education, Vol. 51, No. 12, p. 816, 1974.
Kushner and Hoffman: Synthetic Detergents. Scientific American, Vol. 185, No. 4, p. 26, 1951.
Magliulo: Prostaglandins. Journal of Chemical Education, Vol. 50, No. 9, p. 602, 1973.
Mancott and Tietjen: Polyunsaturation in Food Products. Chemistry, Vol. 47, No. 10, p. 29, 1974.
O'Malley and Schrader: The Receptors of Steroid Hormones. Scientific American, Vol. 234, No. 2, p. 32, 1976.
Pike: Prostaglandins. Scientific American, Vol. 255, No. 5, p. 85, 1971.
Ruchelman: Gas Chromatography: Medical Diagnostic Aid. Chemistry, Vol. 43, No. 11, p. 14, 1970.

Chapter 7. Vitamins and Coenzymes
Dowling: Night Blindness. Scientific American, Vol. 215, No. 4, p. 78, 1966.
Fulkrod: Vitamin C and the Diet of a Student. Journal of Chemical Education, Vol. 49, No. 11, p. 738, 1972.
Hubbard and Kropf: Molecular Isomers in Vision. Scientific American, Vol. 216, No. 6, p. 64, 1967.
Johnson and Williams: Action of Sight upon the Visual Pigment Rhodopsin. Journal of Chemical Education, Vol. 47, p. 736, 1970.
Mellinkoff: Chemical Intervention. Scientific American, Vol. 229, No. 3, p. 102, 1973.
Pauling: Vitamin C and the Common Cold. San Francisco, W. H. Freeman and Company, 1970.
Scrimshaw and Young: The Requirements of Human Nutrition. Scientific American, Vol. 235, No. 3, p. 50, 1976.
Young: Visual Cells. Scientific American, Vol. 233, No. 4, p. 80, 1970.

Chapter 8. Biochemical Energy
Alberty: Thermodynamics of the Hydrolysis of Adenosine Triphosphate. Journal of Chemical Education, Vol. 46, p. 713, 1969.
Changeux: The Control of Biochemical Reactions. Scientific American, Vol. 212, No. 4, p. 36, 1965.
Dickerson: The Structure and History of an Ancient Protein. Scientific American, Vol. 226, No. 4, p. 58, 1972.
Kirschbaum: Biological Oxidations and Energy Conservation. Journal of Chemical Education, Vol. 45, p. 28, 1968.

Chapter 9. Introduction to Metabolism
Clark and Marcker: How Proteins Start. Scientific American, Vol. 218, No. 1, p. 36, 1968.
Grünewald: The Evolution of Proteins. Chemistry, Vol. 41, No. 1, p. 11, 1968.
Horecker: Pathways of Carbohydrate Metabolism and Their Physiological Significance. Journal of Chemical Education, Vol. 42, p. 244, 1965.
Kretchmer: Lactose and Lactase. Scientific American, Vol. 227, No. 4, p. 70, 1972.
Research Reporter: Why the Stomach does Not Digest Itself. Chemistry, Vol. 46, No. 5, p. 20, 1973.

Chapter 10. Carbohydrate Metabolism
Cohen: The Protein Switch of Muscle Contraction. Scientific American, Vol. 233, No. 5, p. 36, 1975.
Gabrielli: Gluconeogenesis: A Teaching Pathway. Journal of Chemical Education, Vol. 53, No. 2, p. 86, 1976.
Govindjee: The Absorption of Light in Photosynthesis. Scientific American, Vol. 231, No. 6, p. 68, 1974.
Horecker: Pathways of Carbohydrate Metabolism and Their Physiological Significance. Journal of Chemical Education, Vol. 42, p. 244, 1965.

Margaria: The Sources of Muscular Energy. Scientific American, Vol. 266, No. 3, p. 84, 1972.
Paston: Cyclic AMP. Scientific American, Vol. 227, No. 2, p. 97, 1972.
Smith and York: Stereochemistry of the Citric Acid Cycle. Journal of Chemical Education, Vol. 47, p. 588, 1970.

Chapter 11. Lipid Metabolism
Benditt: The Origin of Atherosclerosis. Scientific American, Vol. 236, No. 2, p. 74, 1977.
Brady: Hereditary Fat-Metabolism Diseases. Scientific American, Vol. 229, No. 2, p. 88, 1973.
Gibson: The Biosynthesis of Fatty Acids. Journal of Chemical Education, Vol. 42, p. 236, 1965.
Green: The Synthesis of Fat. Scientific American, Vol. 202, No. 2, p. 46, 1960.
Magliulo: Prostaglandins. Journal of Chemical Education, Vol. 50, No. 9, p. 602, 1973.
Spain: Atherosclerosis. Scientific American, Vol. 215, No. 2, p. 48, 1966.

Chapter 12. Protein Metabolism
Clark and Marcker: How Proteins Start. Scientific American, Vol. 218, No. 1, p. 36, 1968.
Engelman and Moore: Neutron-Scattering Studies of the Ribosome. Scientific American, Vol. 235, No. 5, p. 44, 1976.
Grünewald: The Evolution of Proteins. Chemistry, Vol. 41, No. 1, p. 11, 1968.
Harpstead: High-Lysine Corn. Scientific American, Vol. 225, No. 2, p. 34, 1971.
Howe: Amino Acids in Nutrition. Chemical and Engineering News, p. 74, July 23, 1962.
Mayer: The Dimensions of Human Hunger. Scientific American, Vol. 235, No.3, p. 40, 1976.
Roth: Ribonucleic Acid and Protein Synthesis. Journal of Chemical Education, Vol. 38, p. 217, 1961.
Scrimshaw and Behar: Protein Malnutrition in Young Children. Science, Vol. 133, p. 2039, 1961.
Yanofsky: Gene Structure and Protein Structure. Scientific American, Vol. 216, No. 5, p. 80, 1967.

Chapter 13. The Biochemistry of Genetics
Brown: The Isolation of Genes. Scientific American, Vol. 229, No. 2, p. 20, 1973.
Cooper and Lawton: The Development of the Immune System. Scientific American, Vol. 231, No. 5, p. 58, 1974.
Friedmann: Prenatal Diagnosis of Genetic Disease. Scientific American, Vol. 225, No. 5, p. 34, 1971.
Jerne: The Immune System. Scientific American, Vol. 229, No. 1, p. 52, 1973.
Lane: Rabbit Hemoglobin from Frogs' Eggs. Scientific American, Vol. 235, No. 2, p. 60, 1976.
Maniatis and Ptashne: A DNA Operator-Repressor System. Scientific American, Vol. 234, No. 1, p. 64, 1976.
Mazia: The Cell Cycle. Scientific American, Vol. 230, No. 1, p. 54, 1974.
McKusick: The Mapping of Human Chromosomes. Scientific American, Vol. 228, No. 3, p. 34, 1973.
Notkins and Koprowski: How the Immune Response to a Virus Can Cause Disease. Scientific American, Vol. 228, No. 1, p. 22, 1973.
Ptashne and Gilbert: Genetic Repressors. Scientific American, Vol. 222, No. 6, p. 36, 1970.
Ruddle and Kucherlapati: Hybrid Cell and Human Genes. Scientific American, Vol. 231, No. 1, p. 36, 1974.
Stein, Stein, and Kleinsmith: Chromosomal Proteins and Gene Regulation. Scientific American, Vol. 232, No. 2, p. 46, 1975.
Temin: RNA-Directed DNA Synthesis. Scientific American, Vol. 226, No. 1, p. 24, 1972.

Chapter 14. Body Fluids
Edelman: The Structure and Function of Antibodies. Scientific American, Vol. 223, No. 2, p. 34, 1970.
Herron: Simplified Apparatus for Electrophoresis on Paper. Journal of Chemical Education, Vol. 46, p. 527, 1969.
Merrill: The Artificial Kidney. Scientific American, Vol. 205, No. 1, p. 56, 1961.
Smith: The Kidney. Scientific American, Vol. 188, No. 1, p. 40, 1953.
Surgenor: Blood. Scientific American, Vol. 190, No. 2, p. 54, 1954.

Chapter 15. Biochemistry of Drugs
Alexrod: Neurotransmitters. Scientific American, Vol. 230, No. 6, p. 58, 1974.
Barron, Jarvich, and Bunnell: The Hallucinogenic Drugs. Scientific American, Vol. 210, No. 4, p. 29, 1964.
Berelson and Freedman: A Study in Fertility Control. Scientific American, Vol. 210, No. 5, p. 29, 1964.

Bogue: Drugs of the Future. Journal of Chemical Education, Vol. 46, p. 468, 1969.

Braun: The Reversal of Tumor Growth. Scientific American, Vol. 213, No. 5, p. 75, 1965.

Cairns: The Cancer Problem. Scientific American, Vol. 233, No. 5, p. 64, 1975.

Frei and Frereich: Leukemia. Scientific American, Vol. 210, No. 5, p. 88, 1964.

Gates: Analgesic Drugs. Scientific American, Vol. 215, No. 5, p. 131, 1966.

Grinspoon: Marijuana. Scientific American, Vol. 221, No. 6, p. 17, 1969.

Hansch: Drug Research or the Luck of the Draw. Journal of Chemical Education, Vol. 51, No. 6, p. 360, 1974.

Lieber: The Metabolism of Alcohol. Scientific American, Vol. 234, No. 3, p. 25, 1976.

Nares and Strickland: Barbiturates. Chemistry, Vol. 47, No. 3, p. 15, 1974.

Nichols: How Opiates Change Behavior. Scientific American, Vol. 212, No. 2, p. 80, 1965.

Rademacher and Gilde: Chemical Carcinogens. Journal of Chemical Education, Vol. 53, No. 12, p. 757, 1976.

Weeks: Experimental Narcotic Addiction. Scientific American, Vol. 210, No. 3, p. 46, 1964.

GLOSSARY

Absorption–the process by which the end products of digestion are transported from the small intestine to the blood or lymph.

Acetal–the product formed by the reaction of a molecule of an aldehyde with two molecules of an alcohol in the presence of an acid. Acetal linkages are found in polysaccharides and nucleic acids.

Acetyl coenzyme A–an important intermediate compound in the Krebs cycle, in fatty acid oxidation, and in fatty acid synthesis.

Acid-base balance–the maintenance of the normal pH of the blood by the combined action of the lungs, the kidneys, and the body fluid buffers.

Active site–the portion of the enzyme molecule to which the substrate binds.

Adenine–6-amino purine, an essential constituent of DNA and RNA.

Adipose tissue–a storage form of fat usually found under the skin.

Aldose–a simple sugar or monosaccharide containing an aldehyde group.

Alkaptonuria–a genetic abnormality of tyrosine metabolism due to a deficiency of homogentisic acid oxidase in the tissues.

Allopurinol–a purine derivative used in the treatment of gout.

Allosteric enzymes–enzymes located at the beginning of a series of catabolic reactions to control the rate-limiting step in the series.

Alpha helix–a chain of amino acid units wound into a spiral and held together by hydrogen bonds between a carbonyl group of one amino acid and the imino group of an amino acid residue further along the chain.

Amino acid–in general, a compound containing both an amine group and a carboxylic acid group; in particular, alpha amino acids in which the amino group is attached to the same carbon as the carboxyl group; the building blocks of peptides and proteins.

Amino acid pool–a theoretical temporary pool of amino acids available for all phases of protein metabolism.

Amino acid sequence–the sequence by which amino acids are joined together in peptide linkages to form a polypeptide or protein.

Amino group–the NH_2 group.

Amphetamine–a drug that exerts a powerful stimulus action on the central nervous system.

Amphoteric–compounds such as amino acids that ionize as both acids and bases in aqueous solution.

Anabolism–the biosynthetic processes in the body tissues that are active during growth and in the synthesis of important cellular compounds.

Anaerobic pathway–an anaerobic process for the conversion of glucose to pyruvate.

Analgesic–a pain-relieving drug such as aspirin, phenacetin, morphine, or codeine.

Antibiotics–antibacterial or anti-infectious agents such as the sulfa drugs and penicillin.

Anticodon–the triplet of bases on the t-RNA that binds to the site on the m-RNA specific for the amino acid carried by the t-RNA.

Antienzymes–inhibitors of enzyme action in the body.

Antifertility drugs–drugs related to progesterone that act as oral contraceptives by preventing the normal occurrence of ovulation.

Antihistamines–drugs that antagonize the pharmacological actions of histamine and reduce the intensity of allergic reactions.

Antioxidant–a compound that inhibits the auto-oxidation of unsaturated fatty acids and protects against rancidity of fats.

Apoenzyme–the inactive enzyme protein molecule.

Aromatic amino acids–amino acids containing an aromatic ring in their structure.

Ascorbic acid–a water-soluble vitamin whose deficiency in the diet results in scurvy; also know as vitamin C.

202

Aspirin–common name for acetylsalicylic acid, a common antipyretic and analgesic drug.

Asymmetric carbon atom–a carbon atom that has four different groups attached to it.

Atherosclerosis–the deposition of cholesterol plaques in the aorta and lesser arteries.

ATP–adenosine triphosphate, a key compound in the storage of energy and in the coupling of exergonic to endergonic reactions in the cell.

Barbiturates–a class of drugs that act as general depressants of the central nervous system.

Beta-oxidation–the oxidation of fatty acids with the subsequent splitting off of two-carbon fragments in the form of acetyl CoA.

Bicarbonate buffer–the combination of a bicarbonate salt and a carbonic acid; the single most important buffer involved in acid-base balance.

Biuret reaction–a colorimetric reaction for the qualitative and quantitative determination of proteins.

Blood–the most active transport system in the body, consisting of cellular elements suspended in plasma, the liquid medium.

Blood lipids–the mixture of triglycerides, phospholipids, and cholesterol normally found in the blood.

Blood sugar level–the concentration of glucose in the blood before and after meals and in the fasting state; a reflection of the state of carbohydrate metabolism.

Boat form–a conformation form of cyclohexane that is less stable than the chair form in aqueous solution.

Buffer solution–a solution of a weak acid and its conjugate base or a weak base and its conjugate acid that resists large pH changes; essential in acid-base balance in body fluids.

Calciferol–vitamin D_2; formed when the sterol ergosterol is irradiated with ultraviolet light.

Carbamyl phosphate–an activated form of ammonia and carbon dioxide that initiates the urea cycle.

Carbohydrate–an aldehyde or ketone derivative of a polyhydric alcohol.

Carboxyl group–the $-\overset{\overset{\text{O}}{\|}}{\text{C}}-\text{OH}$ group; characteristic functional group of carboxylic acids.

Carcinogen–a material that causes cancer.

Carcinoma–a cancer or tumor involving the epithelial cells of the body.

Carotenes–a group of fat-soluble pigments that may be converted into vitamin A.

Catabolism–the biodegradative processes that result in the formation of metabolites and the production of chemical energy used in cellular reactions.

Catalyst–a component of a reaction that alters the mechanism, lowers the activation energy, and thus increases the rate but is not consumed.

Cation–a positively charged ion in an electrolyte in aqueous solution or in plasma.

Cell–the fundamental unit of living organisms. Also, an electrical device which converts chemical energy into electrical energy.

Cellulose–a polysaccharide occurring in the supporting structure of plants.

Cephalin–a phospholipid found in brain tissue that is involved in the clotting process.

Cerebroside–a glycolipid found in brain and nervous tissue, often containing galactose as the carbohydrate.

Chair form–the most stable conformation form of cyclohexane in aqueous solution.

Chlorophyll–a magnesium-containing derivative of protoporphyrin which plays an important role in the process of photosynthesis in green plants.

Chloroplasts–small pigment particles in plant cells that contain chlorophyll and are active in photosynthesis.

Cholesterol–the most common sterol found in blood, brain, and nervous tissue and in gall-stones.

Cholesterol synthesis–the complex synthesis of cholesterol from two-carbon compounds as simple as the acetyl group in acetyl CoA.

Chromatography–a technique utilizing layers of paper, powdered cellulose, or ion exchange resins to separate compounds of biochemical interest such as amino acids and lipids.

Chromosomes–large DNA molecules in the nucleus of the cell which contain the genes.

Δ^{11} **cis-retinal**–a component of the visual pigment rhodopsin that is formed from vitamin A in the visual cycle.

Citric acid–common name for 3-hydroxy-3-carboxypentanedioic acid, a constituent of citrus fruits commonly used to impart a sour taste to food products and beverages; also a key compound in the citric acid cycle or Krebs cycle.

Code number–a number used to describe the characteristics of a single enzyme in the classification of enzymes.

Codon–the specific site on the m-RNA, consisting of three consecutive bases that bind a particular amino acid.

Coenzyme–an essential cofactor for the action of an enzyme. Usually an organic molecule related to a vitamin.

Coenzyme A–an important coenzyme, consisting of pantothenic acid, ribose, adenine, and phosphoric acid, which is involved in many metabolic reactions in the body.

Coenzyme Q–a quinone compound, which is reduced to the hydroquinone form during the transport of hydrogen in the electron transport system.

Competitive inhibitors–compounds that directly compete with the substrate for the active site on the enzyme surface.

Cortisone–a steroid first used in the treatment of rheumatoid arthritis.

Coupled reactions–an endergonic reaction linked to an exergonic reaction in the cell. ATP is often used to form a common intermediate, which drives the endergonic reaction to completion.

Creatine–methyl guanidoacetic acid; formed from creatine phosphate during muscular contraction and from glycine and arginine in protein anabolism.

Creatine phosphate–a high-energy compound that functions in muscle contraction.

Creatinine–an anhydride of creatine that is an end product of creatine metabolism in muscle tissue.

Cyanocobalamin–a complex molecule known as vitamin B_{12}; used in the treatment of anemias.

Cyclic-3′,5′-AMP–a derivative of adenylic acid that functions in processes of glycogenolysis to provide glucose to the blood during fasting.

Cytochromes–oxidation-reduction pigments which consist of iron-porphyrin complexes attached to a protein molecule, and which play an important role in the electron transport system.

Cytoplasm–the protoplasmic matrix of the cell, in which are embedded the subcellular particles.

Cytosine–2-oxy-4-amino-pyrimidine, an essential constituent of DNA and RNA.

Cytotoxic drugs–drugs that destroy cancer cells in preference to normal cells.

Dark reaction–a series of reactions in photosynthesis that are not dependent on light energy and which function in carbon fixation (the incorporation of carbon into carbohydrates).

Deamination–a general reaction of protein catabolism in which the amino group of an amino acid is split off, with the formation of ammonia and a keto acid.

Denaturation–changes in the conformation of a protein that may result in altered chemical, biological, or physical properties.

Detergents–mixtures of the sodium salts of the sulfuric acid esters of lauryl and cetyl alcohols which readily form suds in hard water.

Diabetes mellitus–an impairment of carbohydrate oxidation ordinarily caused by the lack of insulin and characterized by high blood sugar levels and ketone bodies in the blood and urine.

Differential centrifugation–the centrifugation of a solution at different speeds so that particles of various sizes may be separated.

Digestion–the action of hydrolytic enzymes in digestive fluids to convert large food molecules into smaller end products.

Dipeptide–two amino acids joined together by the peptide linkage.

Disaccharides–sugars composed of two monosaccharides joined together in an acetal linkage.

DNA–deoxyribonucleic acid, an essential component of the nucleus of cells that is responsible for the hereditary characteristics of an organism.

DNA ligases–enzymes that join together the loose ends of DNA strands in gene manipulation studies.

DNA polymerases–enzymes that catalyze the replication of DNA during cell division.

Dopa–dihydroxyphenylalanine; an intermediate in tyrosine metabolism, used in the treatment of Parkinson's disease.

Edema–a swelling of tissues caused by an increased flow of fluid to the interstitial spaces.

Electrolyte balance–a balance of the positively charged cations and negatively charged anions in the plasma.

Electrolytes–compounds which when added to water result in a solution that conducts electricity; also, the cations and anions in the body fluids.

Electron microscope–a microscope that utilizes beams of electrons to visualize particles too small to be seen in a light microscope.

Electron transport system–a series of reactions which transports electrons from a reduced metabolite to oxygen, with the formation of two or three molecules of ATP.

Embden-Meyerhof pathway–an anaerobic process for the conversion of glucose to pyruvate.

Enantiomers–the optically active (+) and (−) forms of a compound such as lactic acid.

Endergonic reactions–reactions that have a + ΔG, or free-energy change, and do not proceed spontaneously to the right without additional energy being supplied.

Endoplasmic reticulum–a network of tubules and vesicles in the cytoplasm, often continuous with the cell membrane or nucleus.

Energy of activation–the energy required to produce the intermediate activated state in an enzyme reaction.

Enzyme–an organic catalyst, protein in nature, which lowers the activation energy so that the rate of a chemical reaction is compatible with the conditions in a living cell.

Enzyme-substrate complex–a transition state in enzyme reactions proposed by Michaelis and Menten.

Epinephrine–a hormone produced by the adrenal medulla that causes glycogenolysis in the liver, with the liberation of glucose into the blood.

Essential amino acids–amino acids essential for growth which cannot be synthesized by the body and must be supplied by dietary protein.

Estrone–a female sex hormone found in the ovarian follicles that is responsible for the development of secondary sexual characteristics at puberty.

Eukaryote–the type of cell found in higher plant and animal organisms containing a nucleus and organelles.

Exergonic reactions–reactions which have a — ΔG, or free-energy change, and occur spontaneously.

Extracellular fluids–plasma, lymph, and tissue fluids.

FADH–reduced flavin adenine dinucleotide; produced in the electron transport system by the combination of FAD and the electrons from a reduced metabolite.

Fat–an ester of fatty acids and glycerol.

Fatty acid–a naturally-occurring straight-chain organic acid that may be saturated or unsaturated and contains an even number of carbon atoms.

Fatty acid oxidation–a series of reactions requiring several enzymes and cofactors, producing acetyl CoA with the stepwise oxidation of the fatty acid molecule two carbons at a time.

Fatty acid synthesis–a series of reactions starting with acetyl CoA reacting with a biotin-enzyme-CO_2 complex to form malonyl CoA, which is built up to a fatty acid molecule.

Fischer projection formula–a straight-chain planar projection representation of carbohydrate structure.

Flavin adenine dinucleotide–a coenzyme that contains riboflavin or vitamin B_2 and functions in many oxidation-reduction reactions in the body.

Fluorouracil–a pyrimidine analog effective in the treatment of carcinoma of the breast and gastrointestinal tract.

Free-energy change–the change in free energy, ΔG, which results when a chemical reaction occurs. For a reaction to proceed spontaneously to the right, the free energy must decrease.

Fructofuranose–the structure of fructose represented as a cyclic compound in the form of the heterocyclic ring furan.

Genes–segments of a chromosome which code for a single polypeptide chain of a protein or an enzyme.

Genetic engineering–the manipulation of genes to change the characteristics of a host organism, to repair genetic defects, or reverse the growth of cancer cells, among other applications.

Germ cells–reproductive cells that are haploid and possess only one set of chromosomes.

Glomerulus–the filtration unit of the kidney, consisting of a mass of capillaries enclosed in Bowman's capsule.

Glucopyranose–the structure of glucose represented as a cyclic compound in the form of the heterocyclic ring pyran.

Glucose–the most important hexose monsaccharide from a nutritional and physiological standpoint; also called dextrose; the sugar circulating in the blood.

Glucose-6-phosphate–a key compound in glucose metabolism which results from the phosphorylation of glucose; the starting material for the Embden-Meyerhof pathway.

Glycerol–common name for 1,2,3-propanetriol, $HOCH_2CHOHCH_2OH$; also called glycerin. An essential constituent of fat; used in many liquid medicine preparations, cosmetics, and lotions.

Glycogen–a polysaccharide that is the storage form of carbohydrate in the animal body.

Glycogenesis–the synthesis of glycogen from glucose by the liver and muscle tissue.

Glycogenolysis–the degradation of glycogen in the liver to form glucose, which is liberated into the bloodstream.

Glycolysis–the processes by which glucose is broken down to compounds such as pyruvic acid and carbon dioxide and chemical energy in the form of ATP.

Golgi complex–a subcellular component consisting of flattened sacs with accompanying vacuoles involved in the transport of material within the cell.

Guanine–2-amino-6-oxy purine, an essential constituent of DNA and RNA.

Half-life–the time required for the decay of half of the nuclei in a radioactive sample. Also, the time required for one-half the mass of a biological compound to be synthesized or degraded in metabolism.

Haworth structure–the representation of carbohydrates as hemiacetal derivatives of the heterocyclic rings pyran and furan.

Hemiacetal–a compound containing a hydroxy (—OH) group and an alkoxy (—OR) group attached to the same carbon atom. Most often prepared by the reaction of an aldehyde with one mole of alcohol. Hemiacetal linkages are found in the Haworth representation of monosaccharides.

Hemoglobin–the respiratory pigment of the red blood cell, composed of four polypeptide chains and four heme molecules.

Henderson-Hasselbalch equation–the relation between pH, pK, and the concentration of the salt and acid forms of a buffer.

Heparin–a mixed polysaccharide that is often used to prevent the clotting of blood samples.

Heroin–the diacetyl ester of morphine; popular with drug addicts.

Heterocyclic amino acids–amino acids containing a heterocyclic ring in their structures.

Hexoses–important monosaccharides containing six carbon atoms.

High-energy compounds–compounds that yield large amounts of free energy on hydrolysis and are often used to link endergonic processes to those that are exergonic.

Holoenzyme–the complete active enzyme molecule, consisting of the apoenzyme plus the prosthetic group or coenzyme.

Hormone–a regulatory substance synthesized in an endocrine gland that exerts its effect on other tissues of the body.

Hydrogen bonding–weak bonds formed between the hydrogen of an —OH, —NH, or —SH group and an electronegative atom such as oxygen, nitrogen, or fluorine. In biochemistry, bonds formed between carbonyl and imino groups of amino acids in protein structure and between imino and nitrogen and carbonyl groups in purines and pyrimidines in DNA.

Hydrolysis–a reaction of a substance with water in which the elements of water are separated; the breakage of a molecule into smaller fragments by the cleavage of one or more covalent bonds; the elements of water are incorporated at each cleavage point as one of the products combines with the hydrogen while the other product combines with the hydroxyl group of the water.

Hyperglycemia–a blood sugar level in excess of the normal fasting level.

Hypoglycemic drugs–orally-administered drugs that reduce the concentration of glucose in the blood.

Inborn errors of metabolism–defects due to a metabolic block in, for example, amino acid catabolism, that involve the deficiency of a different single, specific enzyme whose synthesis is genetically controlled.

Inhibitors–chemical compounds that inhibit the action of enzymes; may be classified as competitive or noncompetitive.

Insulin–a hormone produced by the beta cells of the pancreas that functions in the normal metabolism of glucose to maintain the proper blood sugar level.

Ion exchange resins–insoluble synthetic resins containing acidic or basic groups such as —SO_3H or —OH.

Isoelectric point–the pH at which an electrically charged molecule, such as an amino acid, contains an equal number of positive and negative charges.

Isoenzymes–enzyme protein molecules that are separated into subunits, each having the same activity as the original enzyme.

Ketone bodies–compounds normally produced in the oxidation of fatty acids, acetoacetic and β-hydroxybutyric acids and acetone; found in increased amounts in the blood and urine in abnormal lipid metabolism and in diabetes mellitus.

Ketose–a simple sugar or monosaccharide that contains a ketone group.

K_m–the substrate concentration in moles per liter that will yield the half-maximum rate in an enzyme reaction; known as the Michaelis constant.

Knoop's theory of beta oxidation–Fatty acids are oxidized on the beta carbon with the subsequent splitting off of two-carbon fragments.

Krebs cycle–an aerobic process for the conversion of pyruvate through acetyl CoA to oxaloacetic acid and chemical energy in the form of ATP.

Kwashiorkor–a deficiency disease of very young children who lack adequate protein in their diet.

Lactic acid–$CH_3CH(OH)COOH$; an important acid in muscular contraction and in many energy cycles in the body.

Lactic acid cycle–a series of reactions involved in the formation of lactic acid from blood glucose and liver and muscle glycogen, and the synthesis of liver glycogen from lactic acid.

Lecithin–a phospholipid which is an ester of phosphatidic acid and choline; involved in the transportation of fats in the body.

Light microscope–a microscope that magnifies objects by focusing light beams through its lens system.

Light reaction–reaction involving the conversion of light energy into chemical energy with the assistance of chlorophyll in the process of photosynthesis.

Lineweaver-Burk plot–a double-reciprocal plot of $1/velocity$ versus $1/substrate$ concentration; used to obtain a graphic evaluation of K_m and $v_{max.}$ in an enzyme reaction.

Lipid–a fatty material characterized by the presence of fatty acids or their derivatives and by their solubility in fat solvents.

Lipoprotein–a complex of triglycerides, phospholipids, and cholesterol with plasma proteins for the transportation of the lipids.

Lobry de Bruyn-von Eckenstein transformation–the formation of a common enediol when glucose, fructose, or mannose is allowed to stand in weak alkaline solution.

LSD–lysergic acid diethylamide, a potent hallucinogenic drug.

Lysolecithin–a poisonous compound, found in the venom of snakes, that is formed by removal of oleic acid from the central carbon atom of lecithin.

Lysosomes–specialized cellular vacuoles that contain hydrolytic enzymes capable of digesting phagocytosed material.

Marijuana–a mild type of hallucinogenic drug obtained from the hemp plant.

Membrane–a thin, flat structure that separates the contents of a cell, organ, cavity, or tubulation from the surrounding external media.

Metabolism–a combination of all the processes of anabolism and catabolism in the cell.

Michaelis constant–the substrate concentration in moles per liter that will yield the half-maximum rate in an enzyme reaction; designated as K_m.

Mitochondria–small oval-shaped subcellular particles that contain the compounds and enzyme systems for complete metabolic cycles; called the "powerhouses of the cell."

Mutarotation–a change in optical activity of, for example, carbohydrate solutions with change in pH or with time.

Myoglobin–a heme-containing protein from muscle tissue whose three-dimensional pattern was first determined by x-ray diffraction.

NADH–reduced nicotinamide adenine dinucleotide, produced in the electron transport system by the combination of NAD and the electrons from a reduced metabolite.

NADPH–reduced nicotinamide adenine dinucleotide phosphate, an important coenzyme in carbohydrate metabolism.

Nicotinamide adenine dinucleotide–a coenzyme that contains nicotinamide or niacin and functions in many oxidation-reduction reactions in the body.

Nitrogen cycle–the cyclic set of reactions involving the synthesis and breakdown of nitrogen compounds by plants, animals, and bacteria.

Noncompetitive inhibitors–compounds that combine with a group at the active site and cannot be displaced by additional substrate.

Nonesterified fatty acids (NEFA)–free fatty acids in the blood thought to be very active in metabolism.

Nucleolus–a small dense round body included in the nucleus of a cell.

Nucleoproteins–conjugated proteins that are found in the cell nucleus, consisting of proteins with a high content of basic amino acids conjugated with nucleic acids.

Nucleoside–the combination of a purine or pyrimidine base with a ribose molecule.

Nucleotide–the combination of a purine or pyrimidine base with a ribose molecule attached to a phosphoric acid molecule at one of its hydroxyl groups.

Nucleus–the very dense, positively charged central portion of an atom which contains the protons and neutrons. In a cell, an essential subcellular particle which directs the activities of the cell and regulates its hereditary characteristics.

Obesity–a condition in which excessive amounts of fat are stored in the fat depots.

Operon–a unit consisting of structural genes and their operator gene; involved in the synthesis of enzyme proteins by the cell.

Optical activity–the characteristic of rotation of the plane of polarized light by a compound in solution.

Optimum pH–the pH at which the maximum rate of an enzyme reaction occurs.

Optimum temperature–dependent on a balance between the rise in activity with increased temperature and the inactivation of the enzyme by heat. Value around $37°$ C for enzymes in the body.

Orinase–an orally administered blood sugar–lowering drug used by diabetic patients.

Oxidative phosphorylation–the formation of ATP by the transfer of electrons from reduced metabolites to oxygen by means of the electron transport system.

Penicillin–the original antibiotic drug used to treat bacterial infections.

Pentoses–important monosaccharides containing five carbon atoms.

Peptide–a general term used to describe polyamides made from α-amino acids. Depending on the number of amino acids, the terms "dipeptide," "tripeptide," . . . "polypeptide" may be used.

Peptide linkage–an amide linkage between the carboxyl group of one amino acid and the amino group of another, with the splitting out of a molecule of water.

Phenacetin–p-ethoxyacetanilide; used as an antipyretic and analgesic drug.

Phenylketonuria–an inborn error of metabolism characterized by the formation and excretion of large amounts of phenylpyruvic acid in the urine due to a deficiency in the synthesis of phenylalanine hydroxylase.

Phosphogluconate pathway–an alternate pathway of oxidation of carbohydrates starting with glucose-6-phosphate and forming 5- and 7-carbon intermediates.

Phospholipids–lipids that are composed of phosphoric acid, glycerol, fatty acids, and a nitrogen-containing compound; constituents of all cells.

Phosphorylase–a liver enzyme involved in glycogenolysis; splits the 1,4-glucosidic linkages in glycogen.

Photophosphorylation–the formation of ATP in the light reaction of photosynthesis.

Photosynthesis–the formation of carbohydrates in the cells of green plants from carbon dioxide and water in the presence of sunlight and chlorophyll.

pK–the pH at which equal concentrations of the salt and acid forms of a buffer would exist, or equal quantities of two ionized forms of an amino acid.

Plasma–the fluid portion of the blood, including the clotting factors but not the cells; separated from the cells when blood collected in the presence of an anticoagulant is centrifuged.

Plasma proteins–a mixture of more than 20 proteins, consisting of albumin, several globulins, and fibrinogen.

Plasmid–a small DNA molecule that often carries genes for a special activity such as resistance to antibiotics.

Polarized light–light traveling in one direction and in one plane.

Polypeptide–a combination of several amino acids joined by the peptide linkage.

Polysaccharides–complex carbohydrates composed of several monosaccharide molecules joined by an acetal linkage.

Proenzyme–the inactive precursor form of an enzyme synthesized by a cell.

Prokaryote–a very small, simple cell without a nucleus or organelles, found in bacteria.

Prostaglandins–cyclic fatty acids found in seminal plasma and other tissues that function to depress the action of cyclic-3′,5′-AMP, thereby affecting carbohydrate, lipid, and protein metabolism.

Purine–a nine-membered heterocyclic ring containing two urea residues.

Pyridoxal phosphate–a coenzyme which contains pyridine derivatives known as vitamin B_6; functions in reactions in amino acid metabolism.

Pyrimidine–a six-membered heterocyclic ring containing one urea residue.

Rancidity–the development of unpleasant odors and tastes by fats when they are allowed to stand in contact with air at room temperature.

Reducing sugar–a sugar that contains free or potential aldehyde or ketone groups capable of reducing metal ions such as Cu^{+2} or Ag^+.

Renal threshold–the concentration of a substance in the blood above which it is excreted in the urine.

Repressor–a specific protein molecule that combines with its operator gene to block the function of the structural genes in protein synthesis.

Respiration–the transportation of oxygen and carbon dioxide by hemoglobin with the assistance of body buffers.

Retinal–an aldehyde of vitamin A that is a key compound in the visual cycle.

Ribosomes–very small, round subcellular nucleoprotein particles associated with the endoplasmic reticulum that serve as the site for the synthesis of protein.

RNA–ribonucleic acid, an essential constituent of all cells; responsible for protein synthesis in the cell.

m-RNA–messenger RNA; synthesized under the direction of DNA in the nucleus; carries information to the ribosomes to direct the sequence of alignment of amino acids in protein synthesis.

t-RNA–transfer RNA; combines with a specific amino acid in the cytoplasm and transfers it to the proper position on the m-RNA on the ribosome.

RNA polymerase–an enzyme that functions in the transcription of DNA.

Saponification–the base-catalyzed hydrolysis of an ester; frequently used to refer to the hydrolysis of a fat by an alkali to produce glycerol and soap.

Serum–the fluid portion that separates from clotted blood.

Sickle-cell anemia–a genetic disease caused by the production of an abnormal hemoglobin molecule, hemoglobin S, which causes sickling of the red blood cells.

Soaps–metallic salts of fatty acids formed by saponification of fat.

Somatic cells–cells that are diploid and possess two sets of chromosomes.

Standard free-energy change–the change in free energy, $\Delta G°$, of a chemical reaction measured under standard conditions of temperature, pressure, and concentration.

Starch–a polysaccharide that is the storage form of carbohydrate in plants.

Steroids–derivatives of cyclic alcohols of high molecular weight that contain the sterol nucleus.

Substrate–the compound acted upon by an enzyme to form end products of the reaction.

Sulfa drugs–antibacterial drugs derived from sulfanilamide.

Testosterone–a male sex hormone produced by the interstitial cells of the testes; responsible for the development of masculine sexual characteristics at puberty.

Tetrahydrofolic acid–a coenzyme that is the reduced form of the vitamin folic acid; functions as a carrier of one-carbon units in the synthesis of purines and methyl groups.

Thiamine pyrophosphate–a coenzyme containing thiamine, or vitamin B_1; functions in the conversion of pyruvic acid to acetaldehyde and with enzymes such as α-keto acid oxidase.

Thymine–5-methyl-2,4-dioxypyrimidine, an essential constituent of DNA.

Tocopherol–one of a group of compounds existing in the alpha, beta, and gamma forms that have vitamin E activity and serve as potent antioxidant compounds.

Tranquilizers–drugs such as phenothiazines that relieve nervousness, anxiety symptoms, inability to sleep, and neurotic behavior.

Transamination–a mechanism for the conversion of keto acids to amino acids, often employing glutamic or aspartic acids as the source of the amino group.

Transcription–the process by which DNA in the nucleus transcribes the information to m-RNA for the synthesis of a specific protein.

Translation–the specific programming of the amino acids on the m-RNA molecules on the ribosomes to synthesize a protein containing a definite sequence of amino acids.

Triglyceride–a molecule of fat composed of glycerol and three molecules of fatty acid joined by ester linkages.

Triglyceride synthesis–a series of reactions in which activated forms of glycerol and fatty acids combine to form triglycerides.

Unsaturated fatty acids–straight-chain organic acids with one or more double bonds in the hydrocarbon chain.

Uracil–2,4-dioxypyrimidine, an essential constituent of RNA.

Urea–$H_2NC(O)NH_2$, a by-product of many reactions in the body. Used synthetically in a condensation reaction in the preparation of barbiturates.

Urea cycle–a series of reactions in the liver that starts with CO_2 and NH_3 and forms carbamyl phosphate and eventually urea as an end product.

Uric acid–an oxidized purine molecule that is formed from nucleosides and represents the end product of purine metabolism in man.

Urine formation–the process beginning with the production of a protein-free filtrate of the blood in the glomerulus of the kidney, followed by selective reabsorption by the tubules of substances needed by the body.

Villi–finger-like projections on the inner surface of the small intestine that greatly increase the absorbing surface.

Vitamin–an organic compound essential in the diet in trace amounts to prevent deficiency diseases such as beriberi and scurvy.

Water balance–the daily balance between the fluid intake and fluid excretion by the body.

Zwitterions–substances such as amino acids that ionize as both acids and bases in aqueous solutions.

INDEX

211